INDEX

Acknowledgements

I would like to express my gratitude to the many people who saw me through this book; to all those who provided support, talked things over, read, wrote, offered comments, allowed me to quote their remarks and assisted in the editing, proofreading and design.

I would like to thank my students who encouraged me to publish this book. Above all I want to thank my wife Rashmi, my kids Shriya & Sushant and the rest of my family, who supported and encouraged me in spite of all the time it took me away from them. It was a long and difficult journey for them.

I would like to thank Prof. (Dr.) Satyajit Chakrabari, Vice Chancellor of University of Engineering and Management, Jaipur, who inspired me throughout this process of selection and editing. I would also like to thank my mentor Prof. Dr. Shishir Chandra Bhaduri, Principal, Maharishi Arvind Institute of Technology, who is a constant torch bearer of my life.

Last and not least: I beg forgiveness of all those who have been with me over the course of the years and whose names I have failed to mention."

INTRODUCTION TO VECTOR CALCULUS

CLASSIFICATION OF VECTORS

Broadly vectors can be classified into two categories–

(i) **Axial Vectors :** Where a vector has rotational motion lying along the normal to the plane of rotation of the body and remains unchanged under inversion. e.g.: Torque, angular momentum etc.

(ii) **Polar Vector :** Where a vector has linear motion in a particular direction but changes under inversion or reflection. e.g. displacement, position vector, velocity etc.

Some special vectors

(i) **Unit vector :** It is a vector with unit magnitude and characterizes the direction of the vector mathematically it is denoted by

$$\hat{A} = \frac{\vec{A}}{|A|}$$

In Cartesian coordinate system, let us choose three unit vectors along three mutually perpendicular axes as $\hat{i}, \hat{j}$ and $\hat{k}$ in x, y and z directions respectively. Then any arbitrary vector $\vec{A}$ can be expressed as

$$\vec{A} = A_x\hat{i} + A_y\hat{j} + A_z\hat{k}$$

where A_x, A_y and A_z are called the components of $\vec{A}$ in x, y and z directions.

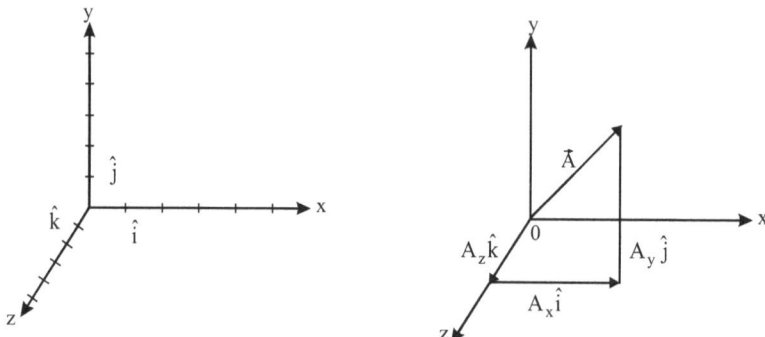

The magnitude of $\vec{A}$ is given using parallelogram law

$$|\vec{A}| = \sqrt{A_x^2 + A_y^2 + A_z^2}$$

hence unit vector along $\vec{A}$ is given by

$$\hat{A} = \frac{A_x\hat{i} + A_y\hat{j} + A_z\hat{k}}{\sqrt{A_x^2 + A_y^2 + A_z^2}}$$

Direction cosines

The cosines of the angles, which $\vec{A}$ makes with x, y and z axis are called direction cosines of the vector. If l, m and n are the direction cosines along $\overrightarrow{ox}$, $\overrightarrow{oy}$ and $\overrightarrow{oz}$ axes, then

$$l = \cos\alpha = \frac{A_x}{|\vec{A}|} \qquad \text{or} \qquad A_x = l|\vec{A}|$$

$$m = \cos\beta = \frac{A_y}{|\vec{A}|} \qquad \text{or} \qquad A_y = m|\vec{A}|$$

$$n = \cos\gamma = \frac{A_z}{|\vec{A}|} \qquad \text{or} \qquad A_z = n|\vec{A}|$$

and

$$l^2 + m^2 + n^2 = \frac{A_x^2 + A_y^2 + A_z^2}{|\vec{A}|^2} = \frac{|\vec{A}|^2}{|\vec{A}|^2} = 1$$

so $$l^2 + m^2 + n^2 = \cos^2 \alpha + \cos^2 \beta + \cos^2 \gamma = 1$$

and $$\vec{A} = |\vec{A}| \left(l\hat{i} + m\hat{j} + n\hat{k} \right)$$

and the unit vector

$$\hat{a} = \frac{\vec{A}}{|\vec{A}|} = l\hat{i} + m\hat{j} + n\hat{k}$$

(ii) **Null Vector :** Any vector with magnitude zero is called null vector. It is collinear with every vector and denoted by $\vec{O}$.

(iii) **Collinear or parallel vector :** When vectors are parallel, then these are collinear vectors, whatsoever their magnitudes may be. Direction of there vectors may be some or opposite.

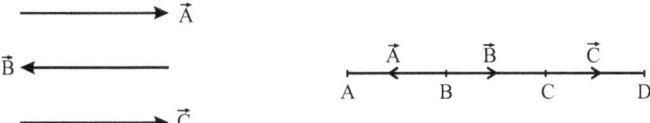

When any scalar is multiplied to any vector then the resultant vector becomes collinear with original one. e.g. $\vec{B} = \lambda\vec{A}$ i.e. $\vec{B}$ vector is λ times $\vec{A}$ with same direction as of $\vec{A}$.

(iv) **Coplanar vectors :** When vectors lies in the same geometrical plane they are called coplanar vectors. Otherwise these are called non-coplanar vectors.

(v) **Like vectors :** The collinear vectors with same sense of direction irrespective of magnitude are called like vectors.

(vi) **Reciprocal vectors :** When the magnitude of a vector is reciprocal to the magnitude of other vector with same direction then it is called reciprocal vector. It is written

as $\dfrac{1}{\vec{A}}$ i.e. $\vec{A}^{-1} = \dfrac{\hat{a}}{|\vec{A}|}$ where $\hat{a}$ is the unit vector along the direction of $\vec{A}$.

Product of vectors

(i) **Scalar product or dot product :**

When the result of product of two vectors is a scalar quantity then this product is known as scalar (or dot) product of the given vectors.

Mathematically it is obtained by multiplying the magnitudes of the vectors with cosines of the angle between them

i.e. $$\vec{A}.\vec{B} = AB\cos\theta$$

$$\vec{A}.\vec{B} = |\vec{A}||\vec{B}|\cos\theta$$

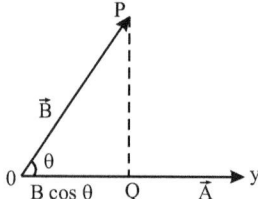

Alternatively scalar product may be defined as multiplication of one vector with component of another in the direction of first.

In case of Cartesian unit vectors

$$\hat{i}.\hat{i} = i.i\cos 0 = 1.1 = 1 = \hat{j}.\hat{j} = \hat{k}.\hat{k}$$

and $$\hat{i}.\hat{j} = ij\cos 90° = 0 = \hat{j}.\hat{k} = \hat{i}.\hat{k}$$

if $$\vec{A} = A_x\,\hat{i} + A_y\,\hat{j} + A_z\,\hat{k}$$

and $$\vec{B} = B_x\,\hat{i} + B_y\,\hat{j} + B_z\,\hat{k}$$

then $$\vec{A}.\vec{B} = A_x\,B_x + A_y\,B_y + A_z\,B_z$$

(a)   Scalar product obeys commutative law i.e. $\vec{A}.\vec{B} = \vec{B}.\vec{A}$

(b)   It obeys distributive law i.e. $\vec{A}.(\vec{B}+\vec{C}) = \vec{A}.\vec{B} + \vec{A}.\vec{C}$

(c)   Two non-zero vectors are orthogonal or perpendicular when $\theta = 90°$ i.e. $\cos\theta = 0$

then $$\vec{A}.\vec{B} = 0$$

similarly two vectors are collinear when $\theta = 0$ or π i.e. $\cos\theta = \pm 1$

then for $\theta = 0$, $$\vec{A}.\vec{B} = AB$$

and for $\theta = \pi$, $$\vec{A}.\vec{B} = -AB$$

Physical examples–

(i)   Work done $W = \vec{F}.\vec{ds}$

(ii)   Power $= \vec{F}.\vec{v}$

(iii)   Magnetic flux of a magnetic field $= \vec{B}.\overrightarrow{ds}$ where $\vec{B}$ is magnetic flux density over an area $\overrightarrow{ds}$.

(iv)   Electric flux of an electric field $= \vec{E}.\overrightarrow{ds}$ where $\vec{E}$ the electric field intensity through elementary area $\overrightarrow{ds}$.

(ii)   Vector product or cross product

When the product of two vectors is a vector quantity, then the product is called vector product or cross product mathematically it is written as

$$\vec{C} = \vec{A} \times \vec{B}$$

$$= \left|\vec{A}\right|\left|\vec{B}\right| \sin \theta. \, \hat{n} \quad \text{where} \ \ 0 \le \theta \le \pi$$

here $\hat{n}$ is the unit vector in the direction of normal to the plane containing $\vec{A}$ and $\vec{B}$ such that $\vec{A}, \vec{B}$ and $\vec{C}$ from a right handed coordinate system with rotation from $\vec{A}$ to $\vec{B}$.

for $\hat{i}, \hat{j}$ and $\hat{k}$.

$$\hat{i} \times \hat{i} = \text{i i} \sin 0 = 0 = \hat{j} \times \hat{j} = \hat{k} \times \hat{k}$$

and

$$\hat{i} \times \hat{j} = \text{i j} \sin 90 = 1 = \hat{j} \times \hat{k} = \hat{k} \times \hat{i}$$

but

$$\hat{j} \times \hat{i} = \text{j i} \sin(-90) = -1 = \hat{k} \times \hat{j} = \hat{i} \times \hat{k}$$

if

$$\vec{A} = A_x \hat{i} + A_y \hat{j} + A_z \hat{k}$$

and

$$\vec{B} = B_x \hat{i} + B_y \hat{j} + B_z \hat{k}$$

then

$$\vec{A} \times \vec{B} = \left|\vec{A}\right|\left|\vec{B}\right| \sin \theta \, \hat{n}$$

$$= \begin{vmatrix} \hat{i} & \hat{j} & \hat{k} \\ A_x & A_y & A_z \\ B_x & B_y & B_z \end{vmatrix}$$

or

$$\vec{A} \times \vec{B} = \left(A_y B_z - A_z B_y\right)\hat{i} - \left(A_x B_z - A_z B_x\right)\hat{j} + \left(A_x B_y - A_y B_x\right)$$

and
$$\hat{n} = \frac{\vec{A} \times \vec{B}}{\left| \vec{A} \times \vec{B} \right|}$$

hence
$$\sin \theta = \frac{\left| \vec{A} \times \vec{B} \right|}{\left| \vec{A} \right| \left| \vec{B} \right|}$$

where
$$\left| \vec{A} \right| = \sqrt{A_x^2 + A_y^2 + A_z^2}$$

and
$$\left| \vec{B} \right| = \sqrt{B_x^2 + B_y^2 + B_z^2}$$

if the rotation from $\vec{A}$ to $\vec{B}$ is anti clockwise then $\vec{C} = \vec{A} \times \vec{B}$ is +ve. and if rotation is clockwise then $\vec{C} = \vec{A} \times \vec{B}$ is −ve.

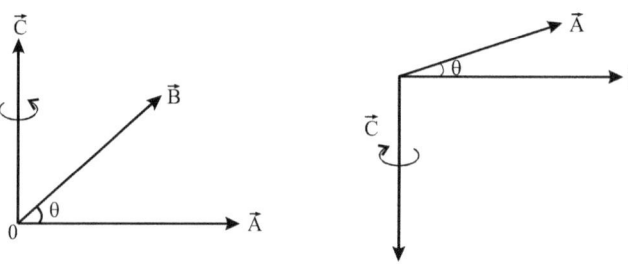

Properties

(a)   Cross product is not commutative

$\vec{A} \times \vec{B} \neq -\vec{B} \times \vec{A}$ but $\vec{A} \times \vec{B} = -\vec{B} \times \vec{A}$

(b)   It is distributive i.e. $\vec{A} \times \left(\vec{B} + \vec{C} \right) = \vec{A} \times \vec{B} + \vec{A} \times \vec{C}$

(c)   If two vectors are collinear or parallel then
$$\theta = 0 \text{ or } \pi$$
then $\sin \theta = 0$

and $\vec{A} \times \vec{B} = 0$

(d)   Two vectors are perpendicular then $\theta = 90°$

so $\vec{A} \times \vec{B} = \left| \vec{A} \right| \left| \vec{B} \right| \hat{n}$

Some examples

(i) Moment of forces

$$\vec{\tau} = \vec{r} \times \vec{F}$$

(ii) Angular momentum $\vec{L} = \vec{r} \times \vec{p} = m(\vec{r} \times \vec{v})$

(iii) Linear velocity $\vec{v} = \vec{\omega} \times \vec{r}$

(iv) force on a charged particle

$$\vec{F} = q(\vec{v} \times \vec{B}) \text{ where q is in coulombs.}$$

(v) Force on a charged particle moving through electric and magnetic field is

$$\vec{F} = q(\vec{E} + \vec{v} \times \vec{B}). \text{ this is known as Lorentz force.}$$

Scalar Triple product

When a vector is scalarly multiplied with the cross product of other two vectors then the result is called scalar triple product.

$$\left[\vec{A}\,\vec{B}\,\vec{C}\right] = \vec{A}.\left(\vec{B} \times \vec{C}\right) = \vec{B}.\left(\vec{C} \times \vec{A}\right) = \vec{C}.\left(\vec{A} \times \vec{B}\right)$$

$$\vec{A}.\left(\vec{B} \times \vec{C}\right) = \begin{vmatrix} A_x & A_y & A_z \\ B_x & B_y & B_z \\ C_x & C_y & C_z \end{vmatrix}$$

Vector triple product

When any vector is vectorily multiplied with vector product of other two vectors taken in cyclic order then the result is known as vector triple product.

$$\vec{A} \times \left(\vec{B} \times \vec{C}\right) = \vec{B}\left(\vec{A}.\vec{C}\right) - \vec{C}\left(\vec{A}.\vec{B}\right)$$

$$\vec{B} \times \left(\vec{C} \times \vec{A}\right) = \vec{C}\left(\vec{B}.\vec{A}\right) - \vec{A}\left(\vec{B}.\vec{C}\right)$$

$$\vec{C} \times \left(\vec{A} \times \vec{B}\right) = \vec{A}\left(\vec{C}.\vec{B}\right) - \vec{B}\left(\vec{C}.\vec{A}\right)$$

Properties

(i) $\vec{A} \times \left(\vec{B} \times \vec{C}\right) + \vec{B} \times \left(\vec{C} \times \vec{A}\right) + \vec{C} \times \left(\vec{A} \times \vec{B}\right) = 0$

(ii) $\left(\vec{A} \times \vec{B}\right) \times \vec{C} = -\vec{C} \times \left(\vec{A} \times \vec{B}\right) = -\vec{A}\left(\vec{B}.\vec{C}\right) + \vec{B}\left(\vec{A}.\vec{C}\right)$

(iii) $\left(\vec{A} \times \vec{B}\right).\left(\vec{C} \times \vec{D}\right) = \left(\vec{A}.\vec{C}\right)\left(\vec{B}.\vec{D}\right) - \left(\vec{A}.\vec{D}\right)\left(\vec{B}.\vec{C}\right)$

(iv) $\vec{A} \times \left(\vec{B} \times \left(\vec{C} \times \vec{D}\right)\right) = \vec{B}\left(\vec{A}.\left(\vec{C} \times \vec{D}\right)\right) - \left(\vec{A}.\vec{B}\right)\left(\vec{C} \times \vec{D}\right)$

Vector differentiation

This is the limiting value of ratio of a vector to the change of a scalar as the change tends to zero is called vector differentiation.

$$\vec{f}(u + \delta u)$$

$$\vec{\delta f} = \vec{f}(u + \delta u) - \vec{f}(u)$$

$$\vec{f}(u)$$

$$\frac{d\vec{f}}{du} = \lim_{\delta u \to 0} \frac{\delta f}{\delta u}$$

$$= \lim_{\delta u \to 0} \frac{\vec{\delta}(u + \delta u) - \vec{f}(u)}{\delta u}$$

Properties

(a) $\dfrac{d}{du}\left(\vec{A} \pm \vec{B}\right) = \dfrac{d\vec{A}}{du} \pm \dfrac{d\vec{B}}{du}$

(b) $\dfrac{d}{du}\left(\vec{A}.\vec{B}\right) = \dfrac{d\vec{A}}{du}.\vec{B} + \vec{A}.\dfrac{d\vec{B}}{du}$

(c) $\dfrac{d}{du}\left(\vec{A} \times \vec{B}\right) = \vec{A} \times \dfrac{d\vec{B}}{du} + \dfrac{d\vec{A}}{du} \times \vec{B}$

(d) $\dfrac{d}{du}\left(\vec{A}.\left(\vec{B} \times \vec{C}\right)\right) = \vec{A}.\left(\vec{B} \times \dfrac{d\vec{C}}{du}\right) + \vec{A}.\left(\dfrac{d\vec{B}}{du} \times \vec{C}\right) + \dfrac{d\vec{A}}{du}.\left(\vec{B} \times \vec{C}\right)$

(e) $\dfrac{d}{du}\left\{\vec{A} \times \left(\vec{B} \times \vec{C}\right)\right\} = \vec{A} \times \left(\vec{B} \times \dfrac{d\vec{C}}{du}\right) + \vec{A} \times \left(\dfrac{d\vec{B}}{du} \times \vec{C}\right) + \dfrac{d\vec{A}}{du} \times \left(\vec{B} \times \vec{C}\right)$

(f) $\dfrac{d}{dt}\vec{a}(s) = \dfrac{d\vec{a}}{ds}\dfrac{ds}{dt}$

(g) $\quad \dfrac{d}{dt}(\vec{e}.\vec{e}) = \vec{e}.\dfrac{d\vec{e}}{dt} + \dfrac{d\vec{e}}{dt} \cdot \vec{e} = 2\vec{e} \cdot \dfrac{d\vec{e}}{dt}$

or $\quad \dfrac{d}{dt}(\vec{e}^2) = 2\vec{e}.\dfrac{d\vec{e}}{dt}$

(h) $\quad \dfrac{d}{dt}\{\phi\,\vec{a}(t)\} = \phi\dfrac{d\vec{a}}{dt}$

Fields

A field is a region in space, where a physical quantity can be specified at every point of the region. Fields may be classified as either scalar or vector depending upon the type of function involved i.e. if a scalar function is taken care of, the field is called a scalar field or if a vector function is involved then that is called a vector field. The temperature of the atmosphere, the height of the surface of earth above sea level are examples of scalar fields. The wind velocity, the gravity force on a mass in space or the force on a charged body in an electric field are examples of vectors fields.

Vector differential operator

The vector differential operator (del operator) is given by

$$\vec{\nabla} = \frac{\partial}{\partial x}\,\hat{i} + \frac{\partial}{\partial y}\,\hat{j} + \frac{\partial}{\partial z}\,\hat{k}$$

It remains invariant under rotation of coordinate system.

Directional Derivatives

Suppose $f(x, y, z)$ and $f(x + \Delta x, y + \Delta y, z + \Delta z)$ are two scalar point functions and

$$df = \frac{\partial f}{\partial x}\,dx + \frac{\partial f}{\partial y}\,dy + \frac{\partial f}{\partial z}\,dz$$

Variation of f can be written as

$$\frac{df}{dr} = \frac{\partial f}{\partial x}\frac{dx}{dr} + \frac{\partial f}{\partial y}\frac{dy}{dr} + \frac{\partial f}{\partial z}\frac{dz}{dr}$$

$$= \left(\frac{\partial f}{\partial x}\hat{i} + \frac{\partial f}{\partial y}\hat{j} + \frac{\partial f}{\partial z}\hat{k}\right).\left(\frac{dx}{dr}\hat{i} + \frac{dy}{dr}\hat{j} + \frac{dz}{dr}\hat{k}\right)$$

$$\frac{df}{dr} = \vec{\nabla}f.\hat{b}$$

Where unit vector $\qquad \hat{b} = \dfrac{dx}{dr}\hat{i} + \dfrac{dy}{dr}\hat{j} + \dfrac{dz}{dr}\hat{k}$

Hence the directional derivative is

$$\boxed{\frac{df}{dr} = \vec{\nabla} f \cdot \frac{\vec{a}}{|\vec{a}|}}$$ where f can be a vector function also .

Coordinates systems

(i)   Cartesian coordinates system

Any point P(x, y, z) can be represented as below–

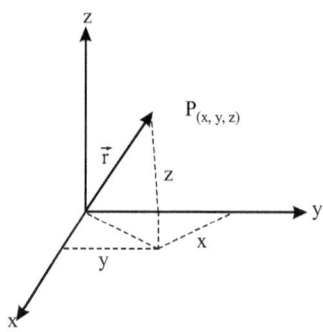   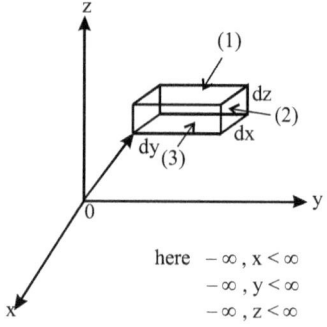

here $-\infty , x < \infty$
$-\infty , y < \infty$
$-\infty , z < \infty$

The position vector for P can be

$$\vec{r} = x\hat{i} + y\hat{j} + z\hat{k}$$

and increase in $\vec{r}$

$$d\vec{r} = dx\,\hat{i} + dy\,\hat{j} + dz\,\hat{k}$$

corresponding volume element

$$dV = dx\,dy\,dz.$$

Surface areas of different sections are

$$ds_1 = dy\,dx\,\hat{k}$$

$$ds_2 = dx\,dz\,\hat{j}$$

$$ds_3 = dy\,dz\,\hat{i}$$

and surface area of areas opposite to these surfaces

$$ds_1' = -dy\,dx\,\hat{k}$$

$$ds_2' = -dx\,dz\,\hat{j}$$

$$ds_3' = -dy\,dz\,\hat{i}$$

(ii)   Cylindrical coordinate system

Here any point is represented in term of cylindrical coordinate (ρ, ϕ, z) as shown.

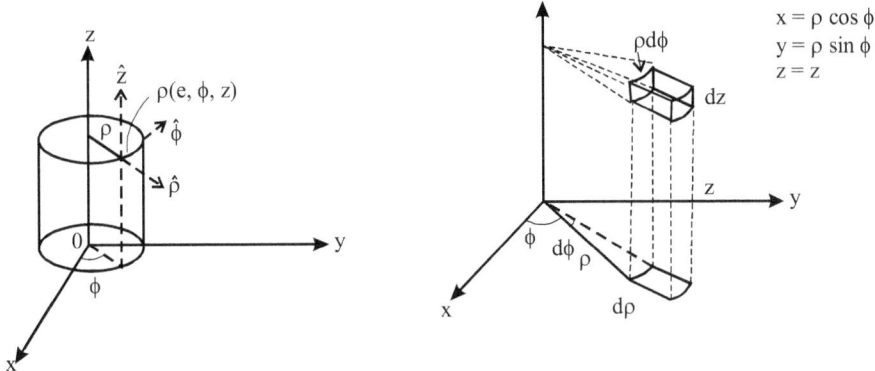

$$x = \rho \cos \phi$$
$$y = \rho \sin \phi$$
$$z = z$$

Any vector in cylindrical coordinate system can be written as

$$\vec{A} = A_\rho \, \hat{\rho} + A_\phi \hat{\phi} + A_z \, \hat{z}$$

differential length i.e. increase in length is given by

$$d\vec{l} = d\rho \, \hat{\rho} + \rho \, d\phi \, \hat{\phi} + dz \, \hat{z}$$

and

$$dl = \sqrt{d\rho^2 + (\rho d\phi)^2 + (dz)^2}$$

differential surfaces are

$$ds_1 = \rho \, d\phi \, dz \, \hat{\rho}$$

$$ds_2 = dz \, d\rho \, \hat{\phi}$$

$$ds_3 = \rho \, d\phi \, d\rho \, \hat{z}$$

Differential volume is

$$dV = d\rho \, \rho d\phi \, dz$$

$$= \rho \, d\rho \, d\phi \, dz$$

(iii)   Spherical coordinates system

In this system any point at the surface of a sphere can be represented in terms of spherical coordinates (r, θ, ϕ) as shown.

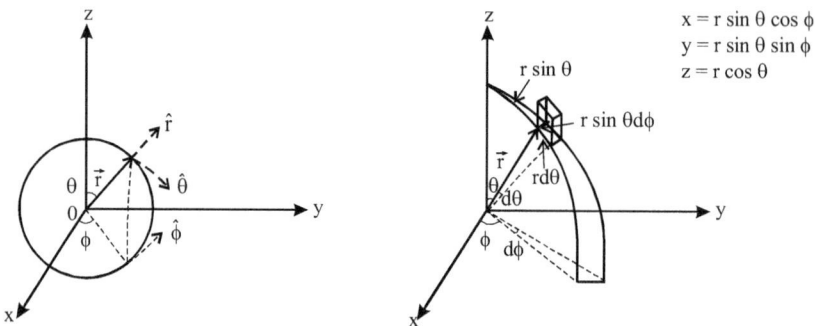

Here r is the radius of imaginary sphere with $0 \le r \le \infty$.

$0 \le \theta \le \pi$ and $0 \le \phi \le 2\pi$.

Here differential length $\quad \overrightarrow{dl} = A_r \hat{r} + A_\theta \hat{\theta} + A_\phi \hat{\phi}$

$$\overrightarrow{dl} = dr\,\hat{r} + rd\theta\,\hat{\theta} + r\sin\theta d\phi\,\hat{\phi}$$

differential surfaces are

$$ds_1 = rd\theta\,\hat{\theta} \times r\sin\theta d\phi\,\hat{\phi} = r^2 \sin\theta\,d\theta\,d\phi\,\hat{r}$$

$$ds_2 = r\sin\theta d\phi\,\hat{\phi} \times dr\,\hat{r} = r\sin\theta\,d\phi\,dr\,\hat{\theta}$$

$$ds_3 = dr\,\hat{r} \times rd\theta\,\hat{\theta} = rdr\,d\theta\,\hat{\phi}$$

and differential volume is

$$dV = dr(rd\theta)(r\sin\theta\,d\phi)$$

$$= r^2 \sin\theta\,d\theta\,dr\,d\phi$$

The Gradient of a scalar point function

When del operator $(\vec{\nabla})$ operates on a scalar point function f(x, y, z), the obtained vector is called gradient $(\vec{\nabla}f)$ of a scalar.

$$\vec{\nabla}f = \text{grad } f = \left(\frac{\partial}{\partial x}\hat{i} + \frac{\partial}{\partial y}\hat{j} + \frac{\partial}{\partial z}\hat{k}\right)f(x,y,z)$$

so,

$$\vec{\nabla}f = \frac{\partial f}{\partial x}\hat{i} + \frac{\partial f}{\partial y}\hat{j} + \frac{\partial f}{\partial z}\hat{k}$$

Gradient in Cylindrical Coordinate System

here $\qquad$ $\psi(r) = \psi(\rho, \phi, z)$ $\qquad$ ...(1)

so $\qquad$ $d\psi = \dfrac{\partial \psi}{\partial \rho}\, d\rho + \dfrac{\partial \psi}{\partial \phi}\, d\phi + \dfrac{\partial \psi}{\partial z}\, dz$ $\qquad$ ...(2)

and $\qquad$ $d\psi = d\vec{r}\ \text{grad}\ \psi$ $\qquad$ ...(3)

But the line element in cylindrical coordinates system

$$d\vec{r} = d\rho\,\hat{\rho} + \rho\,d\phi\,\hat{\phi} + dz\,\hat{z} \qquad ...(4)$$

and any vector

$$\vec{A}(r) = \vec{A}(\rho, \phi, z) = A_\rho\,\hat{\rho} + A_\phi\,\hat{\phi} + A_z\,\hat{z} \qquad ...(5)$$

so $\qquad$ $\text{grad}\ \psi = (\text{grad}\ \psi)_\rho\,\hat{\rho} + (\text{grad}\ \psi)_\phi\,\hat{\phi} + (\text{grad}\ \psi)_z\,\hat{z}$ $\qquad$ ...(6)

so $\qquad$ $d\psi = d\vec{r}\,.\,\text{grad}\ \psi$

$$= (\text{grad}\ \psi)_\rho\, d\rho + (\text{grad}\ \psi)_\phi\, \rho\,d\phi + (\text{grad}\ \psi)_z\, dz \qquad ...(7)$$

comparing (2) and (7)

$$\frac{\partial \psi}{\partial \rho} = (\text{grad}\ \psi)_\rho\,;\, \frac{\partial \psi}{\partial \phi} = \rho(\text{grad}\ \psi)_\phi\,;\, \frac{\partial \psi}{\partial z} = (\text{grad}\ \psi)_z$$

so from (6)

$$\text{grad}\ \psi = \vec{\nabla}\psi = \frac{\partial \psi}{\partial \rho}\,\hat{\rho} + \frac{1}{\rho}\frac{\partial \psi}{\partial \phi}\,\hat{\phi} + \frac{\partial \psi}{\partial z}\,\hat{z}$$

So del operator in cylindrical system is

$$\vec{\nabla} \equiv \frac{\partial}{\partial \rho}\,\hat{\rho} + \frac{1}{\rho}\frac{\partial}{\partial \phi}\,\hat{\phi} + \frac{\partial}{\partial z}\,\hat{z}$$

Gradient in spherical coordinate system

here $\qquad$ $\psi(r) = \psi(r, \theta, \phi)$

$$d\psi = \frac{\partial \psi}{\partial r}\, dr + \frac{\partial \psi}{\partial \theta}\, d\theta + \frac{\partial \psi}{\partial \phi}\, d\phi \qquad ...(1)$$

again $\qquad$ $d\psi = d\vec{r}\,.\,\text{grad}\ \psi$ $\qquad$ ...(2)

line element in spherical system

$$d\vec{r} = dr\,\hat{r} + rd\theta\,\hat{\theta} + r\sin\theta\,d\phi\,\hat{\phi} \qquad ...(3)$$

Any vector in spherical system

$$\vec{A}(r) = \vec{A}(r, \theta, \phi) = A_r\,\hat{r} + A_\theta\,\hat{\theta} + A_\phi\,\hat{\phi} \qquad ...(4)$$

$$\text{grad}\,\psi = (\text{grad}\,\psi)_r\,\hat{r} + (\text{grad}\,\psi)_\theta\,\hat{\theta} + (\text{grad}\,\phi)_\phi\,\hat{\phi} \qquad ...(5)$$

as

$$d\psi = d\vec{r}\,.\,\text{grad}\,\psi$$

$$= (\text{grad}\,\psi)_r\,dr + (\text{grad}\,\psi)_\theta\,rd\theta + (\text{grad}\,\psi)_\phi\,r\sin\theta\,d\phi \,...(6)$$

comparing equation (1) and (6)

$$(\text{grad}\,\psi)_r = \frac{\partial\psi}{\partial r};\,(\text{grad}\,\psi)_\theta = \frac{1}{r}\frac{\partial\psi}{\partial\phi}$$

and

$$(\text{grad}\,\psi)_\phi = \frac{1}{r\sin\theta}\frac{\partial\psi}{\partial\phi}$$

so

$$\text{grad}\,\psi = \vec{\nabla}\psi = \frac{\partial\psi}{\partial r}\,\hat{r} + \frac{1}{r}\frac{\partial\psi}{\partial\theta}\,\hat{\theta} + \frac{1}{r\sin\theta}\frac{\partial\psi}{\partial\phi} \qquad ...(7)$$

so del operator in spherical coordinates

$$\vec{\nabla} \equiv \frac{\partial}{\partial r}\,\hat{r} + \frac{1}{r}\frac{\partial}{\partial\theta}\,\hat{\theta} + \frac{1}{r\sin\theta}\frac{\partial}{\partial\phi}$$

Properties of Gradient

(i) $\nabla(u + v) = \nabla v + \nabla u$

(ii) $\nabla(vu) = v\nabla u + u\nabla v$

(iii) $\nabla\left(\dfrac{u}{v}\right) = \dfrac{v\nabla u - u\nabla v}{v^2}$

(iv) $\nabla v^n = nv^{n-1}\,\nabla v$

(v)   if $\vec{A} = \nabla T$, T is scalar potential of $\vec{A}$.

(vi)   Line integrals of gradients are path independent.

(vii)   If $\nabla X = 0$, it refers to a maximum or minimum point of the function X.

(viii) A vector field derived from the gradient of a scalar field is called as Lameller field.

(ix) Grad V or $\vec{\nabla}V$ points in the direction of the maximum rate of change in V.

(x) the projection of $\left(\vec{\nabla}V\right)$ in the direction of a unit vector is $\vec{\nabla}V.\vec{a}$ and called directional derivative of V along $\vec{a}$.

Divergence of a vector (at a point on the vector field)

When del operator is operated as dot product on a differentiable vector field, the obtained scalar is called the divergence of the vector.

$$\text{div } \vec{A} = \vec{\nabla}.\vec{A} = \left(\frac{\partial}{\partial x}\hat{i} + \frac{\partial}{\partial y}\hat{j} + \frac{\partial}{\partial z}\hat{k}\right).\left(A_x\hat{i} + A_y\hat{j} + A_z\hat{k}\right)$$

$$\vec{\nabla}.\vec{A} = \frac{\partial A_x}{\partial x} + \frac{\partial A_y}{\partial y} + \frac{\partial A_z}{\partial z} \qquad ...(1)$$

Divergence may be defined as the limiting values of the ratio of the flux of a vector across any closed surface around the point to the volume of the enclosure, when volume is contracted to zero.

i.e.
$$\vec{\nabla}.\vec{A} = \lim_{\Delta V \to 0} \frac{\oint_s \vec{A}.\vec{ds}}{\Delta V} \qquad ...(2)$$

Hence divergences of $\vec{A}$ is the net outward flux of the vector field $\vec{A}$ per unit volume as the volume shrinks to zero. Mathematically divergence is a measure of how much quantity comes out from some point.

The divergence of a vector field can also be viewed as the limit of the field's source strength per unit volume, which is positive at a point of source, negative at a point of sink and zero at where no source or sink.

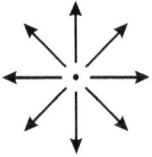

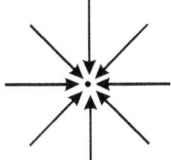

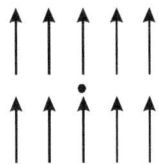

Source → positive   Sink → negative   Zero divergence
divergence   divergence

If $\vec{\nabla} \cdot \vec{v} = 0$ it implies physically that there is no inflow or outflow and the vector is called solenoidal vector. The equation $\vec{\nabla} \cdot \vec{A} = 0$ is also known as continuity equation of an incompressible liquid.

Gauss Divergence Theorem

It states that the total outward flux of a vector field $\vec{A}$ through the closed surface S is equal to the volume integral of the divergence of $\vec{A}$ over the enclosed volume.

$$\oint \vec{A} \cdot \vec{ds} = \int_V \vec{\nabla} \cdot \vec{A} \, dV$$

i.e. $\oint$ flow out through the surface $= \int$ faucets within the volume.

It related the surface integral of a vector $\vec{A}$ with the volume integral of its divergence.

Divergence in cylindrical coordinate system

as
$$\text{div } \vec{A} = \vec{\nabla} \cdot \vec{A}$$

$$= \left(\frac{\partial}{\partial \rho} \hat{\rho} + \frac{1}{\rho} \frac{\partial}{\partial \phi} \hat{\phi} + \frac{\partial}{\partial z} \hat{z} \right) \cdot \left(A_\rho \hat{\rho} + A_\phi \hat{\phi} + A_z \hat{z} \right)$$

$$= \frac{\partial A_\rho}{\partial \rho} + \frac{1}{\rho} \left(\frac{\partial \hat{\rho}}{\partial \phi} A_\rho + \frac{\partial A_\rho}{\partial \phi} \hat{\rho} + \frac{\partial \hat{\phi}}{\partial \phi} A_\phi + \hat{\phi} \frac{\partial A_\theta}{\partial \phi} + \hat{z} \frac{\partial A_z}{\partial \phi} \right) + \frac{\partial A_z}{\partial z} \quad ...(1)$$

But
$$\hat{\phi} = \frac{\partial \hat{\rho}}{\partial \phi}$$

and
$$-\hat{\rho} = \frac{\partial \hat{\phi}}{\partial \phi}$$

so
$$\vec{\nabla} \cdot \vec{A} = \frac{\partial A_\rho}{\partial \rho} + \frac{\hat{\phi}}{\rho} \left(\hat{\phi} A_\rho + \hat{\rho} \frac{\partial A_\rho}{\partial \phi} - \hat{\rho} A_\phi + \hat{\phi} \frac{\partial A_\phi}{\partial \phi} + \hat{z} \frac{\partial A_z}{\partial \phi} \right) + \frac{\partial A_z}{\partial z}$$

$$= \frac{\partial A_\rho}{\partial \rho} + \frac{1}{\rho} A_\rho + \frac{1}{\rho} \frac{\partial A_\phi}{\partial \phi} + \frac{\partial A_z}{\partial z}$$

so
$$\vec{\nabla} \cdot \vec{A} = \frac{1}{\rho} \frac{\partial}{\partial \rho} \left(\rho A_\rho \right) + \frac{1}{\rho} \frac{\partial A_\phi}{\partial \phi} + \frac{\partial A_z}{\partial z} \quad ...(2)$$

Divergence in spherical coordinate system

$$\text{div } \vec{A} = \vec{\nabla}.\vec{A}$$

$$= \left(\frac{\partial}{\partial r}\hat{r} + \frac{1}{r}\frac{\partial}{\partial \theta}\hat{\theta} + \frac{1}{r\sin\theta}\frac{\partial}{\partial \phi}\hat{\phi} \right) \cdot \left(\hat{r} A_r + A_\theta \hat{\theta} + A_\phi \hat{\phi} \right)$$

as

$$\hat{\theta} = \frac{\partial \hat{r}}{\partial \theta}; \; \hat{\phi}\sin\theta = \frac{\partial \hat{r}}{\partial \phi}; \; -\hat{r} = \frac{\partial \hat{\theta}}{\partial \theta}$$

$$\hat{\phi}\cos\theta = \frac{\partial \hat{\theta}}{\partial \phi} \quad \text{and} \quad -\hat{r}\sin\theta - \hat{\theta}\cos\theta = \frac{\partial \hat{\phi}}{\partial \phi}$$

$$\boxed{\vec{\nabla}.\vec{A} = \frac{1}{r^2}\frac{\partial}{\partial r}\left(r^2 Ar \right) + \frac{1}{r\sin\theta}\frac{\partial}{\partial \theta}\left(\sin\theta A\theta \right) + \frac{1}{r\sin\theta}\frac{\partial A_\phi}{\partial \phi}}$$

Properties of divergences

(i)   It produces a scalar field.

(ii)   The divergence of a scalar field makes no sense.

(iii) $\vec{\nabla}.\left(\vec{A} + \vec{B} \right) = \vec{\nabla}.\vec{A} + \vec{\nabla}.\vec{B}$

(iv)   The vector field with zero divergence is called solenoid field.

Curl of a vector (rotation of a vector)

When del operator operates on a differentiable vector field with cross product, the obtained vector is called curl of the vector.

If $\vec{A}$ is a differentiable vector point function.

Then
$$\vec{\nabla} \times \vec{A} = \text{curl } \vec{A} = \begin{vmatrix} \hat{i} & \hat{j} & \hat{k} \\ \frac{\partial}{\partial x} & \frac{\partial}{\partial y} & \frac{\partial}{\partial z} \\ A_1 & A_2 & A_3 \end{vmatrix}$$

$$\vec{\nabla} \times \vec{A} = \left(\frac{\partial A_3}{\partial y} - \frac{\partial A_2}{\partial z} \right)\hat{i} - \left(\frac{\partial A_3}{\partial x} - \frac{\partial A_1}{\partial z} \right)\hat{j} + \left(\frac{\partial A_2}{\partial x} - \frac{\partial A_1}{\partial y} \right)\hat{k}$$

Physically it is the maximum circulation of $\vec{A}$ per unit area as the area tends to zero and whose direction is the normal direction of the area when the area is oriented to make the circulation maximum. i.e. it is the amount of maximum line integral at any point in a vector field per unit area around a closed curve.

So
$$\vec{V} \times \vec{A} = \lim_{\Delta s \to 0} \frac{\oint \vec{A}.\overline{dl}}{\Delta s}$$

In hydrodynamics curl of a vector field at point indicates the amount of rotation of that vector about that point.

If $\vec{V} \times \vec{A} = 0$ i.e. it is irrotational motion of incompressible fluid otherwise irrotational motion of compressible fluid.

If the field is not irrotational then it is called vortex field.

Divergence of a curl is always zero. $\vec{V}.(\vec{V} \times \vec{A}) = 0$.

Curl of a gradient is always zero $\vec{V} \times \vec{V}A = 0$

Properties

(i)   The curl of a vector is a vector

(ii) $\vec{V} \times (\vec{A} \times \vec{B}) = \vec{A}(\vec{V}.\vec{B}) - \vec{B}(\vec{V}.\vec{A}) + (\vec{B}.\vec{V}) - (\vec{A}.\vec{V})\vec{A}$

(iii) $\vec{V} \times (\vec{A} + \vec{B}) = \vec{V} \times \vec{A} + \vec{V} \times \vec{B}$

(iv) $\vec{V} \times (\vec{A}\,\vec{B}) = \vec{A}\,\vec{V} \times \vec{B} + \vec{V}\,\vec{B} \times \vec{A}$

(v) $\vec{V} \times (\vec{V}\,\vec{B}) = 0$

(vi) $\vec{V}.(\vec{V} \times \vec{B}) = 0$

(vii) $\vec{V} \times (\vec{V} \times \vec{B}) = \vec{V}(\vec{V}.\vec{B}) - \vec{V}^2\vec{B}$

(viii) $\vec{V} \times \vec{B} = 0$ i.e. irrotational or conservative field.

(ix) $\text{curl}(\text{div } \vec{A}) = 0$ $\vec{V} \times (\vec{V} \cdot \vec{A}) = 0$

Stoke's theorem

It states that the line integral of a vector around a closed curve is equal to the surface integral of the curl of that vector over that bounded by that closed curve.

i.e.
$$\oint_c \vec{A}.\overline{dl} = \oint_s (\vec{V} \times \vec{A}).\overline{ds}$$

Stoke's theorem transforms the surface integral of the curl of a vector into the line integral of that vector over the boundary of the surface.

Curl in cylindrical coordinate system

$$\vec{\nabla} \times \vec{A} = \left(\hat{\rho} \frac{\partial}{\partial \rho} + \frac{1}{\rho} \frac{\partial}{\partial \phi} \hat{\phi} + \frac{\partial}{\partial z} \hat{z} \right) \times \left(A_\rho \hat{\rho} + A_\phi \hat{\phi} + A_z \hat{z} \right)$$

$$= \hat{\rho} \times \left[\frac{\partial A_\rho}{\partial \rho} \hat{\rho} + \frac{\partial A_\phi}{\partial \rho} \hat{\phi} + \frac{\partial A_z}{\partial \rho} \hat{z} \right]$$

$$+ \frac{\hat{\phi}}{\rho} \times \left[\frac{\partial \hat{\rho}}{\partial \phi} A_\rho + \frac{\partial A_\phi}{\partial \phi} \hat{\rho} + \frac{\partial \hat{\phi}}{\partial \phi} A_\phi + \frac{\partial A_\phi}{\partial \phi} \hat{\phi} + \frac{\partial A_z}{\partial \phi} \hat{z} \right]$$

$$+ \hat{z} \times \left[\frac{\partial A_\rho}{\partial z} \hat{\rho} + \frac{\partial A_\phi}{\partial z} \hat{\phi} + \frac{\partial A_z}{\partial z} \hat{z} \right]$$

$$= \left[\frac{\partial A_\phi}{\partial \rho} \hat{z} - \frac{\partial A_z}{\partial \rho} \hat{\phi} \right] + \frac{1}{\rho} \left[-\frac{\partial A_\rho}{\partial z} \hat{z} + A_\phi \hat{z} + \frac{\partial A_z}{\partial \phi} \hat{\rho} \right] + \left[\frac{\partial A_\rho}{\partial z} \hat{\phi} - \frac{\partial A_\phi}{\partial z} \hat{\rho} \right]$$

$$= \hat{\rho} \left[\frac{1}{\rho} \frac{\partial A_z}{\partial \phi} - \frac{\partial A_\phi}{\partial z} \right] + \hat{\phi} \left[\frac{\partial A_\rho}{\partial z} - \frac{\partial_z}{\partial \rho} \right] + \hat{z} \left[\frac{\partial A_\phi}{\partial \rho} - \frac{1}{\rho} \frac{\partial A_\phi}{\partial \phi} + \frac{A_\phi}{\rho} \right]$$

So

$$\vec{\nabla} \times \vec{A} = \hat{\rho} \left[\frac{1}{\rho} \frac{\partial A_z}{\partial \phi} - \frac{\partial A_\phi}{\partial z} \right] + \hat{\phi} \left[\frac{\partial A_\rho}{\partial z} - \frac{\partial A_z}{\partial \rho} \right] + \frac{\hat{z}}{\rho} \left[\frac{\partial}{\partial \rho} (\rho A_\phi) - \frac{\partial A_\rho}{\partial \phi} \right]$$

Curl in spherical coordinates

$$\vec{\nabla} \times \vec{A} = \left[\frac{\partial}{\partial r} \hat{r} + \frac{1}{r} \frac{r}{\partial \theta} \hat{\theta} + \frac{1}{r \sin \theta} \frac{\partial}{\partial \phi} \hat{\phi} \right] \times \left[A_r \hat{r} + A_\theta \hat{\theta} + A_\phi \hat{\phi} \right]$$

$$= \frac{\hat{r}}{r \sin \theta} \left[\frac{\partial}{\partial \theta} (\sin \theta \, A_\phi) - \frac{\partial A_\theta}{\partial \phi} \right] + \frac{\hat{\theta}}{r} \left[\frac{1}{\sin \theta} \frac{\partial A_r}{\partial \phi} - \frac{\partial}{\partial r} (r A_\phi) \right] + \frac{\hat{\phi}}{r} \left[\frac{\partial}{\partial r} (r A_\theta) - \frac{\partial A_r}{\partial \theta} \right]$$

The Laplacian

The laplacian of a scalar function A is the divergence of the gradient of A. The gradient of A is a vector and the divergence of vector is a scalar. Hence Laplacian gives a scalar.

Mathematically it is $\vec{\nabla}.\vec{\nabla}A = \nabla^2 A$ ..

In Cartesian coordinates–

$$\nabla^2 A = \vec{\nabla}.\vec{\nabla}A = \left(\frac{\partial}{\partial x} \hat{a}_x + \frac{\partial}{\partial y} \hat{a}_y + \frac{\partial}{\partial z} \hat{a}_z \right) . \left(\frac{\partial A}{\partial x} \hat{a}_x + \frac{\partial A}{\partial y} \hat{a}_y + \frac{\partial A}{\partial z} \hat{a}_z \right)$$

$$\nabla^2 A = \frac{\partial^2 A}{\partial x} + \frac{\partial^2 A}{\partial y} + \frac{\partial^2 A}{\partial z} \qquad \qquad ...(1)$$

In cylindrical coordinates

$$\nabla^2 A = \frac{1}{\rho} \cdot \frac{\partial}{\partial \rho}\left(\rho \frac{\partial A}{\partial \rho}\right) + \frac{1}{\rho^2}\frac{\partial^2 A}{\partial \phi^2} + \frac{\partial^2 A}{\partial z^2} \qquad \qquad ...(2)$$

and in spherical coordinates

$$\nabla^2 A = \frac{1}{r^2} \cdot \frac{\partial}{\partial r}\left[r^2 \frac{\partial A}{\partial r}\right] + \frac{1}{r^2 \sin\theta}\frac{\partial}{\partial \theta}\left(\sin\theta \frac{\partial A}{\partial \theta}\right)$$

$$+ \frac{1}{r^2 \sin^2\theta}\frac{\partial^2 A}{\partial \phi^2} \qquad \qquad ....(3)$$

if $\nabla^2 A = 0$ then the scalar field A is said to be harmonic.

$$\nabla^2 \vec{A} = \vec{\nabla}\left(\vec{\nabla}.\vec{A}\right) - \vec{\nabla}\times\vec{\nabla}\times\vec{A} \text{ if } \vec{A} \text{ is a vector.}$$

$$\nabla^2 \vec{A} = \nabla^2 A_x \hat{a}_x + \nabla^2 A_y \hat{a}_y + \nabla^2 A_z \hat{a}_z .$$

Line Integrals

It is the integral of the tangential component of any vector along the curve.

$$\int_L \vec{A}.\overrightarrow{dl} = \int_a^b |A| \cos\theta \, dl$$

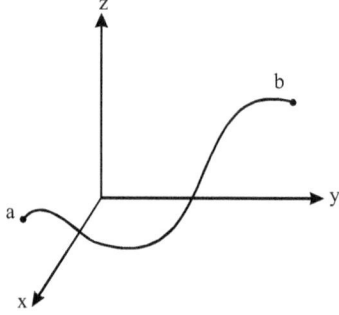

if it is a closed loop then

$$\oint_L \vec{A}.\overrightarrow{dl} = \int_a^b |A| \cos\theta \, dl$$

In Cartesian coordinate system

$$\int_a^b \vec{A}(x,y,z).\,\overrightarrow{dl} = \int_{x_a}^{x_b} A_x\,dx + \int_{y_a}^{y_b} A_y\,dy + \int_{z_a}^{z_b} A_z\,dz$$

in spherical system

$$\int_a^b \vec{A}(r,\,\theta,\,\phi).\,\overrightarrow{dl} = \int_{r_a}^{r_b} A_r\,dr + \int_{\theta_a}^{\theta_b} rA_\theta\,d\theta + \int_{\phi_a}^{\phi_b} r\sin\theta\,A_\phi\,d\phi$$

in cylindrical system

$$\int_a^b \vec{A}(\rho,\,\phi,\,z).\,\overrightarrow{dl} = \int_{\rho_a}^{\rho_b} A_\rho\,d\rho + \int_{\phi_a}^{\phi_b} rA_\phi\,d\phi + \int_{z_a}^{z_b} A_z\,dz$$

Surface integrals

It is the cross product or dot product of a vector $\vec{A}$ with a surfaces element $\overrightarrow{ds}$ for the entire closed surface S.

$$\oint \vec{A}.\overrightarrow{ds} = \oiint_s \vec{A}.\,\overrightarrow{ds} \text{ or } \oint \vec{A} \times \overrightarrow{ds}$$

physically it represents flux of vector $\vec{A}$ over the entire surface s.

e.g. the surfaces integral of an electric field $\vec{E}$ over the surface s represents the flux of $\vec{E}$.

i.e. flux $\phi = \oint_s \vec{E}.\overrightarrow{ds}$

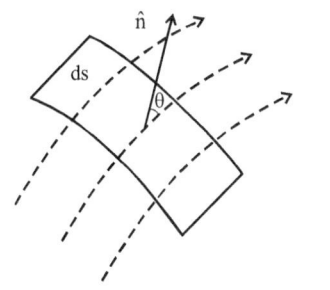
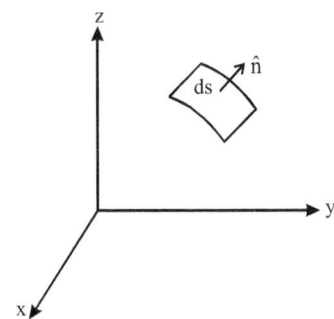

Volume Integral

It is the integral of the scalar function over the entire volume.

$$\int_V \rho \, dV$$ Where ρ is a scalar function and dV is infinitesimal volume

element.

Physical significance depends on the nature of physical quantity for ρ. for example if ρ is density the volume integral gives the total mass and if ρ represents the charge density then volume integral gives total charge bounded by the surfaces.

Some important results involving del operator

(1) $\vec{\nabla}(A+B) = \vec{\nabla}A + \vec{\nabla}B$

(2) $\vec{\nabla} \cdot (\vec{A}+\vec{B}) = \vec{\nabla}.\vec{A} + \vec{\nabla}.\vec{B}$

(3) $\vec{\nabla} \times (\vec{A}+\vec{B}) = \vec{\nabla} \times \vec{A} + \vec{\nabla} \times \vec{B}$

(4) $\vec{\nabla}.(\phi\vec{A}) = \vec{\nabla}\phi.\vec{A} + \phi(\vec{\nabla}.\vec{A})$

(5) $\vec{\nabla} \times (\phi\vec{A}) = \vec{\nabla}\phi \times \vec{A} + \phi(\vec{\nabla} \times \vec{A})$

(6) $\vec{\nabla}.(\vec{A} \times \vec{B}) = \vec{B}.(\vec{\nabla} \times \vec{A}) - \vec{A}.(\vec{\nabla} \times \vec{B})$

(7) $\vec{\nabla} \times (\vec{A} \times \vec{B}) = (\vec{B}.\vec{\nabla})\vec{A} - (\vec{\nabla}.\vec{A})\vec{B} - (\vec{A}.\vec{\nabla})\vec{B} + (\vec{\nabla}.\vec{B})\vec{A}$

(8) $\vec{\nabla}(\vec{A}.\vec{B}) = (\vec{B}.\vec{\nabla})\vec{A} + (\vec{A}.\vec{\nabla})\vec{B} + \vec{B} \times (\vec{\nabla} \times \vec{A}) + \vec{A} \times (\vec{\nabla} \times \vec{B})$

(9) $\vec{\nabla}(\vec{\nabla}\phi) = \nabla^2\phi = \dfrac{\partial^2\phi}{\partial x^2} + \dfrac{\partial^2\phi}{\partial y^2} + \dfrac{\partial^2\phi}{\partial z^2}$

(10) $\vec{\nabla} \times (\vec{\nabla}\phi) = 0$

(11) $\vec{\nabla}.(\vec{\nabla} \times \vec{A}) = 0$

(12) $\vec{\nabla} \times (\vec{\nabla} \times \vec{A}) = \vec{\nabla}(\vec{\nabla}.\vec{A}) - \nabla^2\vec{A}$

Types of vector fields

(i)   Solenoidal and Irrotational field (Lamellar)

if curl $\vec{R} = 0 \Rightarrow \vec{R} = \text{grad } \mu$ where μ is the scalar potential.

$\therefore$ div grad $\mu = \nabla^2 \mu = 0$ (given div $\vec{R} = 0$)

This equation is known as Lapalce's equation and such fields are called Laplacians. e.g. Electrostatic field in free space, gravitational field in free space, thermal fields in equilibrium, magnetostatic fields in current free region, static current field within a linear homogenous isotropic conductor.

(ii)   Irrotational but not solenoidal field

Here curl $\vec{R} = 0$ but div $\vec{R} \neq 0$

again with $\vec{R}$ = grad x, x being the scalar potential but div grad x = $\nabla^2 x \neq 0$

This is called the Poisson's equation and such fields are known as poissonian. e.g. electrostatic fields in a charged medium, electrons inside a thermionic tube, gravitational force inside a mass.

(iii)   Solenoidal but not irrotational field

here   div $\vec{R} = 0$,   but curl $\vec{R} \neq 0$

since curl $\vec{R} \neq 0$ $\vec{R}$ = curl ϕ where ϕ is the vector potential

$\therefore$ div R = div curl ϕ = 0 (A vector identity)

and curl $\vec{R}$ = curl $\phi \neq 0$

$$= \text{grad div } \phi - \nabla^2 \phi \neq 0$$

if now   div $\phi = 0$ then $\nabla^2 \phi \neq 0$

This is similar to Poisson's equation but it is terms of a vector potential. e.g. magnetic field within a conductor carrying a steady current, Rotational motion of an incompressible fluid, time varying electromagnetic field in charge free and current free region.

Neither irrotational nor solenoidal field

for this

$$\text{curl } \vec{R} \neq 0 \quad \text{and div } \vec{R} \neq 0$$
$$\vec{R} = \text{grad x} + \text{curl } \phi$$
$$x = \text{scalar potential}$$
$$\phi = \text{vector potential}$$

$\therefore$ div $\vec{R}$ = div grad x + div curl $\phi \neq 0$

But   div curl $\phi = 0$

so   div grad x $\neq 0$ This is Poisson's equation

and   curl $\vec{R}$ = curl grad x + curl curl ϕ

But curl grad x = 0

$\therefore$ curl curl $\phi \neq 0$

It can be reduced to $\nabla^2 \phi \neq 0$ by assuming div $\phi = 0$. This is the most general type of field.

It can be decomposed into two fields - one is lamellar with x and another solenoidal with ϕ. this is known as 'Helmholz theorem' e.g. Rotational motion of a compressible liquid. Helmholz theorem states that a vector function is determinal uniquely if the values of its curl of divergence are known at all points.

SOLVED EXAMPLES

Q.1 If vector $\vec{A} = 2\hat{i} - \hat{j} + 2\hat{k}$, $\vec{B} = 2\hat{i} - \hat{j}$, $\vec{C} = 2\hat{i} - 3\hat{j} + \hat{k}$ then find

(a) $\vec{A} + \vec{B}$ (b) $\vec{A} - \vec{B}$ (c) $\vec{A} \cdot (\vec{B} \times \vec{C})$ (d) $\vec{B} \cdot (\vec{C} \times \vec{A})$

(e) $\vec{A} \times (\vec{B} \times \vec{C})$ (f) a unit vector perpendicular to both $\vec{B}$ and $\vec{C}$

(g)   Component of $\vec{A}$ along $\vec{B}$.

Solution:

(a)
$$\vec{A} + \vec{B} = \left(2\hat{i} - \hat{j} + 2\hat{k}\right) + \left(2\hat{i} - \hat{j}\right)$$
$$= 4\hat{i} - 2\hat{j} + 2\hat{k}$$

(b)
$$\vec{A} - \vec{B} = \left(2\hat{i} - \hat{j} + 2\hat{b}\right) + \left(2\hat{i} - \hat{j}\right)$$
$$= 2\hat{k}$$

(c)
$$\vec{A} \cdot (\vec{B} \times \vec{C}) = \begin{vmatrix} 2 & -1 & 2 \\ 2 & -1 & 0 \\ 2 & -3 & 1 \end{vmatrix}$$
$$= -2 + 2 - 8 = -8$$

(d)
$$\vec{B} \cdot (\vec{C} \times \vec{A}) = \begin{vmatrix} 2 & -1 & 0 \\ 2 & -3 & 1 \\ 2 & -1 & 2 \end{vmatrix}$$
$$= -10 + 2$$
$$= -8$$

(e)
$$\vec{A} \times (\vec{B} \times \vec{C}) = \vec{B}(\vec{A} \cdot \vec{C}) - \vec{C}(\vec{A} \cdot \vec{B})$$
$$= \left(2\hat{i} - \hat{j}\right)\left[\left(2\hat{i} - \hat{j} + 2\hat{k}\right) \cdot \left(2\hat{i} - 3\hat{j} + \hat{k}\right)\right]$$

$$-\left(2\hat{i}-3\hat{j}+\hat{k}\right)\left[\left(2\hat{i}-\hat{j}+2\hat{k}\right).\left(2\hat{i}-\hat{j}\right)\right]$$

$$=12\hat{i}-3\hat{k}$$

(f)
$$\hat{n} = \frac{\vec{B}\times\vec{C}}{\left|\vec{B}\times\vec{C}\right|} = \frac{\left(2\hat{i}-\hat{j}\right)\times\left(2\hat{i}-3\hat{j}+\hat{k}\right)}{\sqrt{\left(-1\right)^2+\left(-2\right)^2+\left(-4\right)^2}}$$

$$= \frac{-\hat{i}-2\hat{j}-4\hat{b}}{\sqrt{21}}$$

(g)
$$A_{B} = \left|A\right|\cos\theta\,\hat{B}$$

$$= \left(\vec{A}.\hat{B}\right)\hat{B} = \frac{\left(\vec{A}.\vec{B}\right)\vec{B}}{\left|B\right|^2} \quad as \quad \hat{B} = \frac{\vec{B}}{\left|B\right|}$$

$$= \frac{\left[\left(2\hat{i}-\hat{j}+2k\right).\left(2\hat{i}-\hat{j}\right)\right]\left(2\hat{i}-\hat{j}\right)}{\left(2^2+\left(-1\right)^2\right)}$$

$$= \frac{10\hat{i}-5\hat{j}}{5}$$

$$= 2\hat{i}-\hat{j}$$

Q.2: Find the angle between $\vec{A}=2\hat{i}-\hat{j}+\hat{k}$ and $\vec{B}=\hat{i}+\hat{j}+3\hat{k}$.

Solution :

as
$$\vec{A}.\vec{B} = \left|\vec{A}\right|\left|\vec{B}\right|\cos\theta$$

$$\cos\theta = \frac{\vec{A}.\vec{B}}{\left|\vec{A}\right|\left|\vec{B}\right|} = \frac{\left(2\hat{i}-\hat{j}+\hat{k}\right).\left(\hat{i}+\hat{j}+3\hat{k}\right)}{\sqrt{2^2+\left(-1\right)^2+1^2}\,\sqrt{1^2+1^2+3^2}}$$

$$\theta = \cos^{-1}\frac{4}{\sqrt{6}\,\sqrt{11}}$$

$$= 60.5° \text{ approx.}$$

Q.3: If $\vec{A}=2\hat{i}-3\hat{j}+7\hat{k}$ and $\vec{B}=-2\hat{i}+\hat{j}+\hat{k}$, then show that $\vec{A}$ and $\vec{B}$ are perpendicular to each other.

Solution :

$$\vec{A}.\vec{B} = \left(2\hat{i}-3\hat{j}+7\hat{k}\right).\left(-2\hat{i}+\hat{j}+\hat{k}\right)$$

$$= -4 - 3 + 7 = 0$$

so $\qquad \vec{A}.\vec{B} = \left|\vec{A}\right|\left|\vec{B}\right| \cos\theta = 0$

$\Rightarrow \qquad \cos\theta = 0$

$\Rightarrow \qquad \theta = 90°$

so $\vec{A}$ is perpendicular to $\vec{B}$.

Q.4: Find the unit vector perpendicular to both

$\vec{A} = 2\hat{i} + 3\hat{j} - 5\hat{k}$ **and** $\vec{B} = 2\hat{i} + 3\hat{j} - \hat{k}$ **. Also find the angle between them.**

Solution:

As $\qquad \left(\vec{A} \times \vec{B}\right) = \begin{vmatrix} \hat{i} & \hat{j} & \hat{k} \\ 2 & 3 & -5 \\ 2 & 3 & -1 \end{vmatrix}$

$$= 12\hat{i} - 8\hat{j}$$

unit vector perpendicular to both $\vec{A}$ and $\vec{B}$

$$\hat{n} = \frac{\vec{A} \times \vec{B}}{\left|\vec{A} \times \vec{B}\right|}$$

$$= \frac{12\hat{i} - 8\hat{j}}{\sqrt{12^2 + (-8)^2}} = \frac{12\hat{i} - 8\hat{j}}{\sqrt{208}}$$

Again $\qquad \sin\theta = \frac{\left|\vec{A} \times \vec{B}\right|}{\left|\vec{A}\right|\left|\vec{B}\right|}$

$$= \frac{\sqrt{208}}{\sqrt{4+9+25}\,\sqrt{4+9+1}} = \frac{\sqrt{208}}{\sqrt{38}\,\sqrt{14}}$$

$$\theta = \sin^{-1}\left(\frac{\sqrt{208}}{\sqrt{38}\,\sqrt{14}}\right)$$

$$= 38.7° \text{ Approx.}$$

Q.5: A particle is acted upon by two constant forces $\vec{F}_1 = \hat{i} + 4\hat{j} - 3\hat{k}$ **and** $\vec{F}_2 = 3\hat{i} + \hat{j} - \hat{k}$ **due to which particle is displaced from** $\hat{i} + 2\hat{j} + 3\hat{k}$ **to** $4\hat{i} + 5\hat{j} + \hat{k}$ **. Calculate the total work done.**

Solution:

Displacement of the particle

$$\vec{r} = 4\hat{i} + 5\hat{j} + \hat{k} - \left(\hat{i} + 2\hat{j} + 3\hat{k}\right)$$

$$= 3\hat{i} + 3\hat{j} - 2\hat{k}$$

Hence total work done

$$= \text{Total force . displacement}$$

$$= \left(\vec{F}_1 + \vec{F}_2\right).\vec{r}$$

$$= \left[\left(\hat{i} + 4\hat{j} - 3\hat{k}\right) + \left(3\hat{i} + \hat{j} - \hat{k}\right)\right].\left(3\hat{i} + 3\hat{j} - 2\hat{k}\right)$$

$$= \left(4\hat{i} + 5\hat{j} - 4\hat{k}\right).\left(3\hat{i} + 3\hat{j} - 2\hat{k}\right) = 12 + 15 + 8$$

$$= 35 \text{ units.}$$

Q.6: A rigid body is rotating with angular velocity of 5 rad/s about an axis parallel to $3\hat{j} - \hat{k}$ and passing through the point $\hat{i} + \hat{j} - 3\hat{k}$. Find the velocity vector of the particle, when it is at the point $2\hat{i} - 4\hat{j} + \hat{k}$.

Solution : Suppose $\vec{r}$ is the position vector

then

$$\vec{r} = \left(2\hat{i} - 4\hat{j} + \hat{k}\right) - \left(\hat{i} + \hat{j} - 3\hat{k}\right)$$

$$= \hat{i} - 5\hat{j} + 4\hat{k}$$

angular velocity

$$\vec{\omega} = 5 \times \frac{\left(3\hat{j} - \hat{k}\right)}{\left|3\hat{j} - \hat{k}\right|} \frac{5}{\sqrt{10}} \left(3\hat{j} - \hat{k}\right)$$

linear velocity

$$\vec{v} = \vec{\omega} \times \vec{r} = \left(\frac{5}{\sqrt{10}} 3\hat{j} - \hat{k}\right) \times \left(\hat{i} - 5\hat{j} + 4\hat{k}\right)$$

$$= \frac{5}{\sqrt{10}} \begin{vmatrix} \hat{i} & \hat{j} & \hat{k} \\ 0 & 3 & -1 \\ 1 & -5 & 4 \end{vmatrix}$$

$$= \frac{5}{\sqrt{10}} \left(7\hat{i} - \hat{j} - 3\hat{k}\right) \text{ units.}$$

Q.7 : Calculate the torque of a force $-2\hat{i} + 2\hat{j} + 5\hat{k}$ about the point $8\hat{j}$ acting through the point $6\hat{i} + 4\hat{j} + 2\hat{k}$.

Solution : Here

$$\vec{r} = 8\hat{j} - \left(6\hat{i} + 4\hat{j} + 2\hat{k}\right)$$

$$= -6\hat{i} + 4\hat{j} - 2\hat{k}$$

$$\text{torque } \tau = \vec{r} \times \vec{F} = \left(-6\hat{i} + 4\hat{j} - 2\hat{k}\right) \times \left(-2\hat{i} + 2\hat{j} + 5\hat{k}\right)$$

$$= \begin{vmatrix} \hat{i} & \hat{j} & \hat{k} \\ -6 & 4 & -2 \\ -2 & 2 & 5 \end{vmatrix}$$

$$= 24\hat{i} + 34\hat{j} - 4\hat{k}$$

Q.8: A force vector $10\hat{i} + 25\hat{j} + 35\hat{k}$ passes through a point (2, 5, 7). Prove that force is also passing through the origin.

Solution: The position vector

$$\vec{r} = 2\hat{i} + 5\hat{j} + 7\hat{k}$$

and moment of the form about this point i.e. torque

$$\tau = \vec{r} \times \vec{F}$$

$$= \left(2\hat{i} + 5\hat{j} + 7\hat{k}\right) \times \left(10\hat{i} + 25\hat{j} + 35\hat{k}\right)$$

$$= \begin{vmatrix} \hat{i} & \hat{j} & \hat{k} \\ 2 & 5 & 7 \\ 10 & 25 & 35 \end{vmatrix}$$

$$= 0$$

As the moment is zero, which shows that forces is passing through the origin.

Q.9: A force $4\hat{i} - 3\hat{j} + 2\hat{k}$ passes through the point (–9, 2, 1). Find the component of moment of the force about the axis of reference.

Sol.: Here

$$\vec{r} = -9\hat{i} + 2\hat{j} + \hat{k}$$

so moment of force i.e. torque

$$\vec{\tau} = \vec{r} \times \vec{F} = \left(-9\hat{i} + 2\hat{j} + \hat{k}\right) \times \left(4\hat{i} - 3\hat{j} + 2\hat{k}\right)$$

$$= \begin{vmatrix} \hat{i} & \hat{j} & \hat{k} \\ -9 & 2 & 1 \\ 4 & -3 & 2 \end{vmatrix}$$

$$= 7\hat{i} + 22\hat{j} + 19\hat{k}$$

Hence components of moment of force are 7 unit, 22 units and 19 units in x, y and z direction respectively.

Q.10: A proton is moving with velocity 10^8 cm/s along z-axis through an electric field of intensity 3×10^4 volt/cm along x-axis and magnetic field of intensity 2000 gauss along y-axis. Calculate the magnitude and direction of total force.

Solution: Intensity of electric field

$$\vec{E} = 3 \times 10^4 \; \hat{i} \text{ volt} / \text{cm} = 100 \; \hat{i} \text{ esu/cm}$$

$$\text{Proton charge} = 1.6 \times 10^{-19} \text{ C} = 4.8 \times 10^{-10} \text{ esu}$$

$$\text{Magnetic field } \vec{B} = 2000 \; \hat{j} \text{ gauss}$$

$$\text{velocity } \vec{v} = 10^8 \; \hat{k} \text{ cm} / \text{s}$$

so total force acting on the proton

$$\vec{F} = q\left(\vec{E} + \frac{\vec{v} \times \vec{B}}{C}\right)$$

$$= \left(4.8 \times 10^{-10}\right)\left[100\hat{i} + \frac{1}{3 \times 10^{10}}\left\{10^8 \, \hat{k} \times 2000\hat{j}\right\}\right]$$

$$= 4.47 \times 10^{-8} \, \hat{i} \text{ dyne}$$

Hence total force acting on the proton has magnitude $+4.47 \times 10^{-8}$ dyne along the +ve x-direction.

Q.11: Find the value of the constant p so that

$$\vec{A} = 2\hat{i} + \hat{j} - 3\hat{k}, \; \vec{B} = 2\hat{i} - 3\hat{j} - \hat{k} \text{ and } \vec{C} = 3\hat{i} - p\hat{j} + \hat{k} \text{ are coplanar.}$$

Solution: We know that three vectors are said to be coplanar if $\vec{A} \cdot \left(\vec{B} \times \vec{C}\right) = 0$

$$\therefore \qquad \begin{vmatrix} 2 & 1 & -3 \\ 2 & -3 & -1 \\ 3 & -p & 1 \end{vmatrix} = 0$$

$\Rightarrow \qquad\qquad 4p - 38 = 0$

or $\qquad\qquad\quad 4p = 38$

$$p = \frac{38}{4} = 9.5$$

Q.12: Evaluate $\vec{A} \times \left(\vec{B} \times \vec{C} \right)$ where

$$\vec{A} = 2\hat{i} + \hat{j}, \quad \vec{B} = -\hat{i} + \hat{j} + \hat{k} \quad \text{and} \quad \vec{C} = 5\hat{i} - 3\hat{j} + \hat{k}.$$

Solution:

$$\vec{A} \times \left(\vec{B} \times \vec{C} \right) = \vec{B} \left(\vec{A}.\vec{C} \right) - \vec{C} \left(\vec{A}.\vec{B} \right)$$

$$= \left(-\hat{i} + \hat{j} + \hat{k} \right) \left\{ \left(2\hat{i} + \hat{j} + 0\hat{k} \right).\left(5\hat{i} - 3\hat{j} + \hat{k} \right) \right\}$$

$$- \left(5\hat{i} - 3\hat{j} + \hat{k} \right) \left\{ \left(2\hat{i} + \hat{j} + 0\hat{k} \right).\left(-\hat{i} + \hat{j} + \hat{k} \right) \right\}$$

$$= 7 \left(-\hat{i} + \hat{j} + \hat{k} \right) - (1) \left(5\hat{i} - 3\hat{j} + \hat{k} \right)$$

$$= -2\hat{i} + 4\hat{j} + 8\hat{k}$$

Q.13: If $\vec{r}$ si the position vector of any point (x, y, z) and $\vec{A}$ is a constant vector then show that

(i) $\left(\vec{r}.\vec{A} \right).\vec{A} = 0$ is the equation of a constant plane.

(ii) $\left(\vec{r} - \vec{A} \right).\vec{r}$ is the equation of a sphere.

Also show that result of (i) is of the form $Ax + By + Cz + D = 0$ where $D = -\left(A^2 + B^2 + C^2 \right)$ and that of (ii) is of the from $x^2 + y^2 + z^2 = r^2$. [RU 2005]

Solution: (i) Suppose $\vec{A} = (A, B, C)$ and $\vec{r} = (x, y, z)$

$$\left(\vec{r} - \vec{A} \right).\vec{A} = (x - A)A + (y - B)B + (z - C)C$$

$$= xA - A^2 + yB - B^2 + zC - C^2$$

$$= xA + yB + zC - \left(A^2 + B^2 + C^2\right)$$

$$= A_x + B_y + C_z + D$$

where $$D = -\left(A^2 + B^2 + C^2\right)$$

so $$\left(\vec{r} - \vec{A}\right).\vec{A} = 0 \rightarrow A_x + B_y + C_z + D = 0$$

which is an equation of a plane.

(ii) $$\left(\vec{r} - \vec{A}\right).\vec{A} = (x - A)x + (y - B)y + (z - C)z$$

if $$\left(\vec{r} - \vec{A}\right).\vec{A} = 0 \text{ then}$$

$$x^2 + y^2 + z^2 - A_x - B_y - C_z = 0$$

Which is the equation of sphere whose surface touches the origin.

Q.14: A particle moves on the curve $x = 2t^2$, $y = t^2 - 4t$, $z = 3t - 5$ where t is the time. Find the components of velocity and acceleration at time $t = 1$ in the direction $\hat{i} - 3\hat{j} + 2\hat{k}$.

Solution: Position vector

$$\vec{r} = 2t^2\hat{i} + \left(t^2 - 4t\right)\hat{j} + \left(3t - 5\right)\hat{k}$$

so velocity vector

$$\vec{v} = \frac{d\vec{r}}{dt} = \frac{d}{dt}\left(2t^2\right)\hat{i} + \frac{d}{dt}\left(t^2 - 4t\right)\hat{j} + \frac{d}{dt}\left(3t - 5\right)\hat{k}$$

$$= 4t\,\hat{i} + (2t - 4)\hat{j} + 3\hat{k}$$

acceleration $$\vec{a} = \frac{d\vec{v}}{dt}$$

$$= \frac{d}{dt}(4t)\hat{i} + \frac{d}{dt}(2t - 4)\hat{j} + \frac{d}{dt}(3)\hat{k}$$

$$= 4\hat{i} + 2\hat{j} + 0$$

at $$t = 1, \text{ velocity } \vec{v} = 4\hat{i} - 2\hat{j} + 3\hat{k}$$

acceleration $\vec{a} = 4\hat{i} + 2\hat{j}$

and the component of $\vec{v}$ along $\hat{i} - 3\hat{j} + 2\hat{k}$

is $\dfrac{\left(4\hat{i} - 2\hat{j} + 3\hat{k}\right).\left(\hat{i} - 3\hat{j} + 2\hat{k}\right)}{\sqrt{1^2 + (-3)^2 + 2^2}}$

$$= \frac{16}{\sqrt{14}} = \frac{8\sqrt{14}}{7}$$

and component of $\vec{a}$ along $\hat{i} - 3\hat{j} + 2\hat{k}$ is

$$= \frac{\left(4\hat{i} + 2\hat{j}\right).\left(\hat{i} - 3\hat{j} + 2\hat{k}\right)}{\sqrt{1^2 + (-3)^2 + 2^2}} = \frac{-2}{\sqrt{14}}$$

$$= \frac{-\sqrt{14}}{7}$$

Q.15: Calculate the unit vector, which is normal to the surface

$$\phi = x^2 y + xy^2 + 3xyz \text{ at the point } (1, 1, -1).$$

Solution : Here

$$\vec{\nabla}\phi = \left(\frac{\partial}{\partial x}\hat{i} + \frac{\partial}{\partial y}\hat{j} + \frac{\partial}{\partial z}\hat{k}\right)\left(x^2 y + xy^2 + 3xyz\right)$$

$$= \frac{\partial}{\partial x}\left(x^2 y + xy^2 + 3xyz\right)\hat{i} + \frac{\partial}{\partial y}\left(x^2 y + xy^2 + 3xyz\right)\hat{j}$$

$$+ \frac{\partial}{\partial z}\left(x^2 y + xy^2 + 3xyz\right)\hat{k}$$

$$= \left(2xy + y^2 + 3yz\right)\hat{i} + \left(x^2 + 2xy + 3xz\right)j + (3xy)\hat{k}$$

At $(1, 1, -1)$,

$$\vec{\nabla}\phi = (2 + 1 - 3)\hat{i} + (1 + 2 - 3)\hat{j} + 3\hat{k} = 3\hat{k}$$

so the unit vector normal to the surface ϕ at $(1, 1, -1)$ is

$$\frac{3\hat{k}}{\sqrt{3^2}} = \frac{3\hat{k}}{3}$$

$$= \hat{k}$$

Q.16: Find the direction derivative of $\phi(x,y,z) = x^2y + xy^2$ at the point $(2, -1, -4)$ along the direction of the vector $(1, 2, -1)$.

Solution: as

$$\phi = x^2y + xy^2$$

$$\vec{\nabla}\phi = \left(\frac{\partial}{\partial x}\hat{i} + \frac{\partial}{\partial y}\hat{j} + \frac{\partial}{\partial z}\hat{k}\right)\phi(x,y,z)$$

$$= \frac{\partial}{\partial x}\left(x^2y + xy^2\right)\hat{i} + \frac{\partial}{\partial y}\left(x^2y + xy^2\right)\hat{j} + \frac{\partial}{\partial z}\left(x^2y + xy^2\right)\hat{k}$$

$$= \left(2xy + y^2\right)\hat{i} + \left(x^2 + 2xy\right)\hat{j}$$

$$\left(\vec{\nabla}\phi\right)_{(2,-1,-4)} = -3\hat{i}$$

Position vector $\hat{r} = \hat{i} + 2\hat{j} - \hat{k}$

and unit vector along this position vector

$$\hat{n} = \frac{\hat{i} + 2\hat{j} - \hat{k}}{\sqrt{1+4+1}} = \frac{\hat{i} + 2\hat{j} - \hat{k}}{\sqrt{6}}$$

and direction derivative $\quad \vec{\nabla}\phi.\hat{n} = \left(-3\hat{i}\right).\dfrac{\left(\hat{i} + 2\hat{j} - \hat{k}\right)}{\sqrt{6}}$

$$= \frac{-3}{\sqrt{6}} = \frac{-\sqrt{6}}{2}$$

Q.17: Find the equation of the tangent plane and normal line to the surface $2x^2 + y^2 + 2z = 3$ at the point $(2, 1, -3)$.

Solution :

Here $\qquad \phi(x,y,z) = 2x^2 + y^2 + 2z - 3$

$$\therefore \qquad \frac{\partial \phi}{\partial x} = \frac{\partial}{\partial x}\left(2x^2 + y^2 + 2z\right) = 4x$$

$$\frac{\partial \phi}{\partial y} = \frac{\partial}{\partial y}\left(2x^2 + y^2 + 2z\right) = 2y$$

$$\frac{\partial \phi}{\partial z} = \frac{\partial}{\partial z}\left(2x^2 + y^2 + 2z\right) = 2$$

so the components $\dfrac{\partial \phi}{\partial x}, \dfrac{\partial \phi}{\partial y}$ and $\dfrac{\partial \phi}{\partial z}$ at the point (2, 1, –3) will be

$$\frac{\partial \phi}{\partial x} = 4 \times 2 = 8, \quad \frac{\partial \phi}{\partial y} = 2 \times 1 = 2, \quad \frac{\partial \phi}{\partial z} = 2$$

Hence the equation of the tangent plane to the surface at the point (2, 1, –3) is

$$(X-2)8 + (Y-1)2 + (Z+3)2 = 0$$

or $4X + y + Z = 6$

so the equation of normal to the surface at (2, 1, –3) is

$$\frac{X-2}{8} = \frac{Y-1}{2} = \frac{Z+3}{2}$$

or $\dfrac{X-2}{4} = Y - 1 = Z + 3$

Q.18: Find the angle between the surfaces $x^2 + y^2 + z^2 = 9$ and $x^2 + y^2 - z = 3$ at the (1, 2, 2)

Solution:

Suppose $\phi_1 = x^2 + y^2 + z^2$ and $\phi_2 = x^2 + y^2 - z$

so $\vec{\nabla}\phi_1 = 2x\hat{i} + 2y\hat{j} + 2z\hat{k}$

and $\nabla\phi_2 = 2x\hat{i} + 2y\hat{j} - \hat{k}$

and $\left(\vec{\nabla}\phi_1\right)_{1,2,2} = 2\hat{i} + 4\hat{j} + 4\hat{k}$

$$\left(\vec{\nabla}\phi_2\right)_{1,2,2} = 2\hat{i} + 4\hat{j} - \hat{k}$$

since $\vec{\nabla}\phi_1$ and $\vec{\nabla}\phi_2$ are normal to ϕ_1 and ϕ_2

then $\vec{\nabla}\phi_1 . \vec{\nabla}\phi_2 = \left|\vec{\nabla}\phi_1\right|\left|\vec{\nabla}\phi_2\right| \cos\theta$ where θ is the angle between the surfaces ϕ_1 and ϕ_2.

so $\theta = \cos^{-1}\left[\dfrac{\vec{\nabla}\phi_1 . \vec{\nabla}\phi_2}{\left|\vec{\nabla}\phi_1\right|\left|\vec{\nabla}\phi_2\right|}\right]$

$$= \cos^{-1}\frac{4 + 16 - 4}{\sqrt{36 \times 21}} = \cos^{-1}\left(\frac{16}{6\sqrt{21}}\right)$$

$$\theta = 54.41° \text{ approx.}$$

Q.19 (i) Provle that $\vec{P} = \cos\theta_1\,\hat{i} + \sin\theta_1\hat{j}$ and $\cos\theta_2\hat{i} + \sin\theta_2\hat{j}$ are unit vectors in the xy-plane respectively making θ_1 and θ_2 with the x-axis.

(ii) By means of dot product, obtain the formula for $\cos(\theta_2 - \theta_1)$. by similarly formulating P and Q, obtain the formula for $\cos(\theta_2 + \theta_1)$.

(iii) If θ is the angle between P and Q find $\dfrac{1}{2}|P - Q|$ in terms of θ.

Solution:

(i)   Given $\vec{P} = \cos\theta_1\hat{i} + \sin\theta_1\hat{j}$

$$\vec{Q} = \cos\theta_2\hat{i} + \sin\theta_2\hat{j}$$

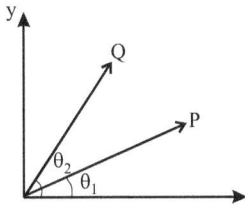

$$|\vec{P}| = \cos^2\theta_1 + \sin^2\theta_1 = 1$$

$$|\vec{Q}| = \cos^2\theta_2 + \sin^2\theta_2 = 1$$

hence $\vec{P}$ and $\vec{Q}$ are unit vectors.

(ii) $\vec{P}.\vec{Q} = |\vec{P}||\vec{Q}| \cos(\theta_2 - \theta_1)$

$$= 1.1 \cos(\theta_2 - \theta_1) \qquad\qquad ...(1)$$

But $\vec{P}.\vec{Q} = \left(\cos\theta_1\hat{i} + \sin\theta_1\hat{j}\right).\left(\cos\theta_2\hat{i} + \sin\theta_2\hat{j}\right)$

$$= \cos\theta_1\cos\theta_2 + \sin\theta_1\sin\theta_2 \qquad\qquad ...(2)$$

so $\cos(\theta_2 - \theta_1) = \cos\theta_1\cos\theta_2 + \sin\theta_1\sin\theta_2$

let $\vec{P}_1 = \vec{P} = \cos\theta_1\hat{i} + \sin\theta_1\hat{j}$

and $\vec{Q}_1 = \cos\theta_2\,\hat{i} - \sin\theta_2\hat{j}$

then $\vec{P}_1 . \vec{Q}_1 = 1.1 \cos(\theta_1 - \theta_2)$

                                      $= \cos\theta_1 \cos\theta_2 - \sin\theta_1 \sin\theta_2$

(iii) $\vec{P}_1$ and $\vec{Q}_1$ are unit vectors.

so $\dfrac{1}{2}(\vec{P}-\vec{Q}) = \dfrac{1}{2}\left| \rho^2 + Q^2 - 2PQ \cos\theta \right|$

                              $= \dfrac{1}{2}\left| 1 + 1 - 2\cos\theta \right|$

                              $= 1 - \cos\theta$

                              $= 2\sin^2\dfrac{\theta}{2}$

Q.20: A vector field is given as $\vec{W} = 4x^2y\,\hat{i} - (7x + 2z)\hat{j} + (4xy + 2z^2)\hat{k}$

(i) **What is the magnitude of the field at point (2. –3, 4).**

(ii) **At what point on z-axis is the magnitude of W equal to unity?** **[RU 2002]**

Solution: (i)

$$\vec{W} = 4x^2y\,\hat{i} - (7x + 2z)\hat{j} + (4xy + 2z^2)\hat{k}$$

at P(2, –3, 4), $\vec{W} = 4(2^2)(-3)\hat{i} - (7\times2 + 4\times2)\hat{j} + (4\times2\times(-3) + 2\times4^2)\hat{k}$

                              $= -48\hat{i} - 22\hat{j} + 8\hat{k}$

$$\left| \vec{W} \right| = \sqrt{48^2 + 22^2 + 8^2} = 53.4$$

(ii)   As the required point is on z-axis so x = 0, y = 0

$$\vec{W} = -2z\hat{j} + 2z^2\hat{k}^2 \text{ for that point.}$$

$\therefore$ $\left| \vec{W} \right| = \sqrt{(-2z)^2 + (2z^2)^2} = \sqrt{4z^2 + 4z^4} = 1$

so $4z^4 + 4z^2 - 1 = 0$

$$z^2 = \dfrac{-4 \pm \sqrt{16 - 16(-1)}}{8} = -\dfrac{1}{2} \pm \sqrt{\dfrac{1}{2}}$$

$$z^2 = -1.207 \text{ and } 0.207$$

taking z as positive $z^2 = 0.207$

$$z = \pm 0.455$$

Q.21: Calculate the differential volume to obtain the expression for volume of the

(i) sphere of radius 'b'

(ii) Semispherical shell of inner radius 'a' and outer radius 'b'.

(iii) Cylinder of radius 'b' and height 'h'.

Solution:

(i) Differential volume in spherical coordinates

$$dV = r^2 \sin\theta\, dr\, d\theta\, d\phi$$

here $\qquad r = 0$ to b, $\qquad \theta = 0$ to π, $\qquad \phi = 0$ to 2π.

So volume of sphere

$$V = \int_V dV = \int_0^b \int_0^\pi \int_0^{2\pi} r^2 \sin\theta\, dr\, d\theta\, d\phi$$

$$= \int_0^b r^2 dr \int_0^\pi \sin\theta\, d\theta \int_0^{2\pi} d\phi$$

$$= 2\pi \int_0^b r^2 dr \left. (-\cos\theta) \right|_0^\pi$$

$$= 2\pi\,(1+1) \int_0^b r^2 dr$$

$$= 4\pi \frac{b^3}{3}$$

so $\qquad V_{sphere} = \dfrac{4\pi}{3} b^3$

(ii) For semispherical shell

$\qquad r_1 = a, \qquad r_2 = b$

so $\qquad dV = r^2 \sin\theta\, dr\, d\theta\, d\phi$

here $\qquad r = a$ to b, $\qquad \theta = 0$ to π, $\phi = 0$ to π

so $\qquad V = \int_a^b \int_0^\pi \int_0^\pi r^2 \sin\theta\, dr\, d\theta\, d\phi$

$$= \int_a^b r^2 dr \int_0^\pi \sin\theta \, d\theta \int_0^\pi d\phi$$

$$= \pi \int_a^b r^2 dr (\cos\theta) \Big|_0^\pi$$

$$= 2\pi \frac{r^2}{3} \Big|_a^b$$

$$= \frac{2\pi}{3} \left(b^3 - a^3 \right)$$

(iii) Differential volume for a cylinder

$$dV = r \, dr \, d\phi \, dz$$

here $r = 0$ to b, $z = 0$ to h, and $\phi = 0$ to 2π.

so $$V = \int_V dV$$

$$V = \int_a^b \int_0^h \int_0^{2\pi} r \, dr \, d\phi \, dz$$

$$= \int_0^b r \, dr \int_0^h dz \int_0^{2\pi} d\phi$$

$$= 2\pi . h . \frac{r^2}{2} \Big|_0^b$$

so $$V_{cyl.} = \pi b^2 h$$

Q.22 : For positive x, y, z let $\rho = 40 \, xyz$ c/m³. **Find the total charge within the region bounded by x = 0, y = 0, $0 \le 2x + 3y \le 10$ and $0 \le z \le 2$.**

Solution : Here

$$Q = \int_0^5 \int_0^y \int_0^z 40 \, xyz \, dx \, dy \, dz$$

$$= \int_0^5 40x \, dx \left[\frac{y^2}{2}\right]_0^{\frac{10-2x}{3}} \left[\frac{z^2}{2}\right]_0^2$$

$$= \frac{40}{9} \int_0^5 \left(100x - 40x^2 + 4x^3\right) dx$$

$$= \frac{40}{9} \left[\frac{100x^2}{2} - \frac{40x^3}{3} - \frac{4x^4}{4}\right]_0^5$$

$$Q = 925.926 \text{ C}$$

Q.23: Given point P in Cartesian coordinate system as P(1, 2, 3). Calculate its coordinates in cylindrical system.

Solution: As given x = 1, y = 2, z = 3

$$\rho = \sqrt{x^2 + y^2} \quad \phi = \tan^{-1}\frac{y}{x}, \quad z = z$$

so

$$\rho = \sqrt{1^2 + 2^2} = \sqrt{5} = 2.236$$

$$\phi = \tan^{-1}\frac{2}{1} = 63.43°$$

$$z = 3$$

so

$$P_{cyl.} = (2.236, 63.43°, 3)$$

Q.24: The cooridnate of a point P in cylindrical system is P(1, 45°, 2). find its equivalent in cartesion system.

Solution:

Here

$$\rho = 1, \quad \phi = 45°, \quad z = 2$$

and

$$x = \rho\cos\phi, \quad y = \rho\sin\phi, \quad z = z$$

$$x = 1.\cos 45 = \frac{1}{\sqrt{2}} = 0.707$$

$$y = 0.707$$

$$z = 2$$

so

$$P_{cart.} = (0.707, 0.707, 2)$$

Q.25: Find the constant m such that the vector

$$\vec{\phi} = (x+3y)\hat{i}+(y-2z)\hat{j}+(x+mz)\hat{k} \text{ is solenoidal.}$$

Solution: The vector will be solenoidal if $\vec{\nabla}.\vec{\phi} = 0$

so $\left(\dfrac{\partial}{\partial x}\hat{i}+\dfrac{\partial}{\partial y}\hat{j}+\dfrac{\partial}{\partial z}\hat{k}\right)\left[(x+3y)\hat{i}+(y-2z)\hat{j}+(x+mz)\hat{k}\right] = 0$

i.e. $\dfrac{\partial}{\partial x}(x+3y)+\dfrac{\partial}{\partial y}(y-2z)+\dfrac{\partial}{\partial z}(x+mz) = 0$

or $1 + 1 + m = 0$

                              $m = -2$

Q.26: Find div $\vec{F}$ and curl $\vec{F}$ if $\vec{F} = \text{grad}\left(x^3+y^3+z^3-3xyz\right).$ [WBUT 2001]

Solution: Here $\vec{F} = \text{grad}\left(x^3+y^3+z^3-3xyz\right)$

$$= \dfrac{\partial}{\partial x}\left(x^3+y^3+z^3-3xyz\right)\hat{i}+\dfrac{\partial}{\partial y}\left(x^3+y^3+z^3-3xyz\right)\hat{j}$$

$$+\dfrac{\partial}{\partial z}\left(x^3+y^3+z^3-3xyz\right)\hat{k}$$

$$= \left(3x^2-3yz\right)\hat{i}+\left(3y^2-3xz\right)\hat{j}+\left(3z^2-3xy\right)\hat{k}$$

$$\text{div }\vec{F} = \dfrac{\partial}{\partial x}\left(3x^2-3yz\right)+\dfrac{\partial}{\partial y}\left(3y^2-3xz\right)+\dfrac{\partial}{\partial z}\left(3z^2-3xy\right)$$

$$= 6(x+y+z)$$

$$\vec{\nabla}\times\vec{F} = \begin{vmatrix} \hat{i} & \hat{j} & \hat{k} \\ \dfrac{\partial}{\partial x} & \dfrac{\partial}{\partial y} & \dfrac{\partial}{\partial z} \\ 3x^2-3yz & 3y^2-3xz & 3z^2-3xy \end{vmatrix}$$

$$= \left(-3x+3x\right)\hat{i}+\left(-3y+3y\right)\hat{j}+\left(-3z+3z\right)\hat{k} = 0$$

Q.27 Show that curl grad f = 0 where f = x²y + 2xy + z².

Solution:

$$\text{grad } f = \frac{\partial f}{\partial x}\hat{i} + \frac{\partial f}{\partial y}\hat{j} + \frac{\partial f}{\partial z}\hat{k}$$

$$= (2xy + 2y)\hat{i} + (x^2 + 2x)\hat{j} + (2z)\hat{k}$$

∴

$$\text{curl grad } f = \begin{vmatrix} \hat{i} & \hat{j} & \hat{k} \\ \dfrac{\partial}{\partial x} & \dfrac{\partial}{\partial y} & \dfrac{\partial}{\partial z} \\ 2xy+2y & x^2+2x & 2z \end{vmatrix}$$

$$= 0 + 0 + (2x + 2 - 2x - 2)\hat{k} = 0$$

Q.28: If the scalar function $\psi(x,y,z) = 2xy + z^2$,. is its corresponding scalar field is solenoidal or irrotational?

Solution : Let

$$\vec{F} = \nabla\psi = 2y\hat{i} + 2x\hat{j} + 2z\hat{k}$$

so

$$\vec{\nabla}.\vec{F} = \frac{\partial}{\partial x}(2y) + \frac{\partial}{\partial y}(2x) + \frac{\partial}{\partial z}(2z)$$

$$= 0 + 0 + 2 = 2$$

$$\neq 0$$

So field is not solenoidal.

Now

$$\vec{\nabla} \times \vec{F} = \begin{vmatrix} \hat{i} & \hat{j} & \hat{k} \\ \dfrac{\partial}{\partial x} & \dfrac{\partial}{\partial y} & \dfrac{\partial}{\partial z} \\ 2y & 2x & 2z \end{vmatrix}$$

$$= 0$$

so field is irrotational.

Q.29: Verify the divergence theorem for the vector function

$$\vec{F} = 4xz\hat{i} - y^2\hat{j} + yz\hat{k}$$

taken over the cube bounded by x = 0, 1 y = 0, 1, z = 0, 1.

[WBUT (math) 2002]

Solution.:

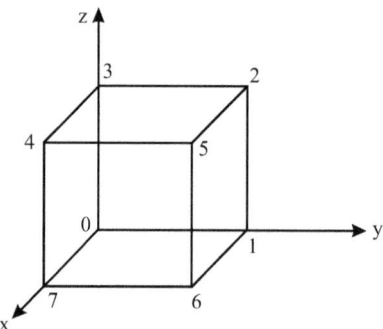

for face 4567; $\hat{n} = \hat{i}$ and $x = 1$

$$\oint_1 \vec{F} . \hat{n} \, ds = \int_0^1 \int_0^1 4z \, dy \, dz = 2$$

$$\oint_2 \vec{F} . \hat{n} \, ds = \left(-\hat{i}\right) dy \, dz = 0 \quad \text{for } 1230$$

for $2561 \, \hat{n} = \hat{j}, \, y = 1$

$$\oint_3 \vec{F} . \hat{n} \, ds = -1$$

and for 3074 $\oint_4 \vec{F} . \hat{n} \, ds = 0$

for face 2345, $\oint_5 \vec{F} . \hat{n} \, ds = \dfrac{1}{2}$

and $\oint_6 \vec{F} . \hat{n} \, ds = 0 \text{ for } 0167$

total $\oint_S \vec{F} . \hat{n} \, ds = 2 + 0 + (-1) + 0 + \dfrac{1}{2} + 0 = \dfrac{3}{2}$

Again $\vec{\nabla} . \vec{F} = \dfrac{\partial}{\partial x}\left(4xz\right) + \dfrac{\partial}{\partial y}\left(-y^2\right) + \dfrac{\partial}{\partial z}\left(yz\right)$

$$= 4z - y$$

$$\therefore \qquad \int_0^1 \int_0^1 \int_0^1 \vec{\nabla}.\vec{F} \, dV \;=\; \int_0^1 \int_0^1 \int_0^1 \left(4z - y\right) dx \, dy \, dz$$

$$= \int_0^1 \int_0^1 \left[\frac{4z^2}{2} - yz\right]_0^1 dx \, dy$$

$$= \frac{3}{2}$$

$$\therefore \qquad \oint_s \vec{F}.\hat{n} \, ds \;=\; \oint_V \vec{\nabla}.\vec{F} \, dV$$

Hence divergence theorem is verified.

Q.30: **Calculate the line integral of** $\vec{A} = \rho \cos\phi \, \hat{\rho} + z \sin\phi \, \hat{z}$ **around the edge L of the wedge defined by** $0 \le \rho \le 4$, $0 \le \phi \le 30°$, $z = 0$.

Solution:

Given $\qquad\qquad\qquad \vec{A} = \rho \cos\phi \, \hat{\rho} + z \sin\phi \, \hat{z}$

differential length

$$\vec{dl} = d\rho \, \hat{\rho} + d\phi \, \hat{\phi} + dz \, \hat{z}$$

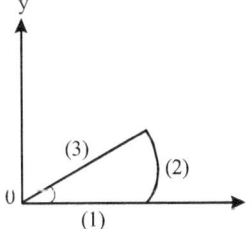

Circulation of $\vec{A}$ around path is

$$\oint_L \vec{A}.\vec{dl} \;=\; \oint_1 \vec{A}.\vec{dl} + \oint_2 \vec{A}.\vec{dl} + \oint_3 \vec{A}.\vec{dl}$$

for (1) $\phi = 0$, $d\phi = 0$, $z = 0$, $dz = 0$

$$\oint_1 \vec{A}.\vec{dl} \;=\; \oint_1 (\rho \cos\phi \hat{\rho} + z \sin\phi) \cdot \left(d\rho \, \hat{\rho} + d\phi \, \hat{\phi} + dz \, \hat{z}\right)$$

$$= \oint \rho \cos\phi \, d\rho = \int_0^4 \rho \, d\rho = \frac{\rho^2}{2}\bigg|_0^4$$

$$= \frac{16}{2} = 8$$

for (2) $\rho = 4$, $d\rho = 0$

$z = 0$, $dz = 0$

$$\oint_2 \vec{A}. \, \vec{dl} = \oint_2 \rho \cos \phi \, d\rho + z \sin \phi \, dz = 0$$

for (3) $\phi = \pi / 6$, $d\phi = 0$, $z = 0$, $dz = 0$

$$\oint_3 \vec{A}. \, \vec{dl} = \oint_3 \rho \cos \phi \, d\rho$$

$$= \int_4^0 \rho \cos \frac{\pi}{6} \, d\rho$$

$$= \frac{\sqrt{3}}{2} \int_4^0 \rho \, d\rho = 0.866 \frac{\rho^2}{2}\bigg|_4^0$$

$$= -6.93$$

So total $\oint_L \vec{A}. \, \vec{dl}$ $= 8 + 0 - 6.93$

$$= 1.07$$

Q.31 Given $\vec{A} = x^2 + xy$, calculate $\int \vec{A}. \, \vec{ds}$ over the region $y = x^2$, $0 < x < 2$.

Solution:

So $y = x^2$, $ds = dx \, dy$

$$\int \vec{A}. \, \vec{ds} = \int \left(x^2 + xy \right) dx \, dy$$

$$= \int\limits_{0}^{2} \int\limits_{y=0}^{x^2} \left(x^2 \, dx \, dy \right) + \int\limits_{0}^{2} \int\limits_{y=0}^{x^2} xy \, dx \, dy$$

$$= \int\limits_{x=0}^{2} x^4 dx + \int\limits_{x=0}^{2} \frac{x^5}{2} \, dx$$

$$= \frac{32}{5} + \frac{64}{12}$$

$$= 11.7$$

Q.32 For a scalar function $\phi = \left[\sin\dfrac{\pi x}{2} + \sin\dfrac{\pi y}{3} \right] e^{-z}$. **Calculate the magnitude direction of maximum rate of increase of ϕ at the point (1, 1, 1).**

Solution: As gradient of a scalar function gives the magnitude and direction of max. rate of change of that

So

$$\nabla \phi = \frac{\partial \phi}{\partial x}\hat{i} + \frac{\partial \phi}{\partial y}\hat{j} + \frac{\partial \phi}{\partial z}\hat{k}$$

$$= \frac{\partial}{\partial x}\left(\sin\frac{\pi x}{2}\sin\frac{\pi y}{3}e^{-z} \right)\hat{i} + \frac{\partial}{\partial y}\left(\sin\frac{\pi x}{2}\sin\frac{\pi y}{3}e^{-z} \right)\hat{j}$$

$$+ \frac{\partial}{\partial z}\left(\sin\frac{\pi x}{2}\sin\frac{\pi y}{3}e^{-z} \right)\hat{k}$$

$$= \left(\frac{\pi}{6}e^{-z}\sin\frac{\pi x}{2} \right)\hat{j} - \left(\sin\frac{\pi x}{2}\sin\frac{\pi y}{3}e^{-z} \right)\hat{k}$$

at (1, 1, 1)

$$(\nabla\phi)_{1,1,1} = \left(\frac{\pi}{6}e^{-1}\sin\frac{\pi}{2} \right)\hat{j} - \left(\sin\frac{\pi}{2}\sin\frac{\pi}{3}e^{-1} \right)\hat{k}$$

$$= \left(\frac{\pi}{6} \times 0.367 \right)\hat{j} - (0.866 \times 0.367)\hat{k}$$

$$= 0.192\,\hat{j} - 0.318\,\hat{k}$$

and

$$|\nabla\phi|_{1,1,1} = \left[(0.192)^2 + (0.318)^2 \right]^{1/2}$$

$$= 0.37$$

and $\hat{j} = \dfrac{\nabla\phi}{|\nabla\phi|} = \dfrac{0.192\hat{j}-0.318\hat{k}}{0.37}$

$$= 0.52\hat{j} - 0.86\hat{k}$$

Q.33 : Determine the divergence of the following vector fields at given points–

(i) $\vec{A} = yz\hat{i} + 4(x+y)\hat{j} + xyz\ \hat{k}$ at $(1,-2,1)$

(ii) $\vec{B} = \rho z \sin \phi\hat{\rho} + 5\rho z \cos\phi\ \hat{\phi} + z\hat{z}$ at $\left[5, \dfrac{\pi}{2}, 1\right]$

(iii) $\vec{C} = 2r\sin \theta \cos\phi\ \hat{r} + \cos\phi\ \hat{\theta} + r\ \hat{\phi}$ at $\left[1, \dfrac{\pi}{3}, \dfrac{\pi}{3}\right]$

Solution: In cartesion

(a) $\vec{\nabla}.\vec{A} = \dfrac{\partial}{\partial x}(yz) + \dfrac{\partial}{\partial y}(4x+4y) + \dfrac{\partial}{\partial z}(xyz)$

$$= 0 + 4 + xy$$

$$\vec{\nabla}.\vec{A} = 4 + xy$$

at $(1,-2,1)$, $\left(\vec{\nabla}.\vec{A}\right)_{1,-2,1} = 4 + 1 \times (-2)$

$$= 2$$

(b)   In cylindrical

$$\vec{\nabla}.\vec{B} = \dfrac{1}{\rho}\dfrac{\partial}{\partial\rho}\left(\rho\ B_\rho\right) + \dfrac{1}{\rho}\dfrac{\partial}{\partial\phi}\left(B_\phi\right) + \dfrac{\partial}{\partial z}B_z$$

given $B_\rho = \rho z \sin \phi_1\quad B_\phi = 5\rho z \cos \phi,\ B_z = z$

$$\vec{\nabla}.\vec{B} = \dfrac{2}{\rho}\rho z \sin\phi + \dfrac{1}{\rho}5.\rho z(-\sin\phi) + 1$$

$$= 1 - 3z \sin \phi$$

at $\left(5, \dfrac{\pi}{2}, 1\right) \Rightarrow$

$$\vec{\nabla}.\vec{B} = 1 - (3 \times 1 \times 1)$$

$$= -2$$

(c)   In spherical system

$$\vec{\nabla}.\vec{C} = \frac{1}{r^2}\frac{\partial}{\partial r}\left(r^2 C_r\right) + \frac{1}{r\sin\theta}\frac{\partial}{\partial\theta}\left(\sin\theta\, C_\theta\right) + \frac{1}{r\sin\theta}\frac{\partial C_\phi}{\partial\phi}$$

$$= \frac{2}{r^2}\sin\theta\cos\phi\,\frac{\partial}{\partial r}\left(r^3\right) + \frac{\cos\phi}{r\sin\theta}\frac{\partial}{\partial\theta}\sin\theta\,\frac{\partial}{\partial\phi}\frac{\partial}{r\sin\theta\,\partial\phi}$$

$$= 6\sin\theta\,\cos\phi + \frac{\cos\phi\cot\theta}{r}$$

at $\left(1, \dfrac{\pi}{3}, \dfrac{\pi}{3}\right)$

$$\vec{\nabla}.\vec{C} = 6\sin\frac{\pi}{3}\cos\frac{\pi}{3} + \frac{\cos\dfrac{\pi}{3}\cot\dfrac{\pi}{3}}{1}$$

$$= 2.6 + 0.288$$

$$= 2.88$$

Q.34: Find the nature of the vector $\vec{F} = 30\hat{i} + 2xy\,\hat{j} + 5xz^2\hat{k}$.   [RU 2003]

Solution :

$$\vec{\nabla}.\vec{F} = \frac{\partial}{\partial x}\left(30\right) + \frac{\partial}{\partial y}\left(2xy\right) + \frac{\partial}{\partial y}\left(5xz^2\right)$$

$$= 2x + 10xz$$

$$\neq 0$$

$$\text{Curl }\vec{F} = \vec{\nabla}\times\vec{F} = \begin{vmatrix} \hat{i} & \hat{j} & \hat{k} \\ \dfrac{\partial}{\partial x} & \dfrac{\partial}{\partial y} & \dfrac{\partial}{\partial z} \\ 30 & 2xy & 5xz^2 \end{vmatrix}$$

$$= 5z^2\,\hat{j} + 2y\hat{k}$$

$$\neq 0$$

as $\vec{\nabla}.\vec{F} \neq 0 \Rightarrow$ fields is not solenoidal

and $\vec{\nabla}\times\vec{F} \neq 0 \Rightarrow$ field is rotational.

Q.35: Given the vector field $\vec{G} = (16xy - 3)\hat{i} + 8x^2\hat{j} - x\hat{k}$.

(i)   Is G irrotational (or conservative)?

(ii)   Find the net flux of G over the cube 0<x, y, z < 1.

(iii)   Determine the circulation of G around the edge of the square z = 0, 0 < x, y < 1. Assume anticlockwise direction.   [RU 2003]

Solution:

(i)
$$\vec{\nabla} \times \vec{G} = \begin{vmatrix} \hat{i} & \hat{j} & \hat{k} \\ \dfrac{\partial}{\partial x} & \dfrac{\partial}{\partial y} & \dfrac{\partial}{\partial z} \\ 16xy - z & 8x^2 & -x \end{vmatrix}$$

$$= 0\hat{i} + (-1+1)\hat{j} + (16x - 16x)\hat{k} = 0$$

So $\vec{G}$ is irrotational.

(ii)   Net flux of $\vec{G}$ over the cube

$$= \int_V (\vec{\nabla} \cdot \vec{G}) dV$$

$$\vec{\nabla} \cdot \vec{G} = \frac{\partial}{\partial x}(16xy - z) + \frac{\partial}{\partial y}(8x^2) + \frac{\partial}{\partial z}(-x)$$

$$= 16y + 0 + 0 = 16y$$

so
$$\int_V (\vec{\nabla} \cdot \vec{G}) dV = \iiint 16y\, dx\, dy\, dz = 16 \int_0^1 dx \int_0^1 dz \int_0^1 y\, dy$$

$$= 16.1.1 \left.\frac{y^2}{2}\right|_0^1 = 8$$

(iii)
$$\oint \vec{G} \cdot \vec{dl} = \int_0^1 (16xy - z)dx \Big|_{\substack{y=0 \\ z=0}}^{x=0} + \int_0^1 8x^2 dy \Big|_{\substack{x=1 \\ y=0 \\ z=0}}^{x=1}$$

$$+ \int_1^0 (16xy - z)dx \Big|_{\substack{y=1 \\ z=0}}^{x=1} + \int_1^0 8x^2 dy \Big|_{\substack{x=0 \\ y=1 \\ z=0}}^{x=0}$$

$$= 0 + 8(1)y \Big|_0^1 + 16(1)\left.\frac{x^2}{2}\right|_1^0 + 0$$

$$= 8 - 8 = 0.$$

SUMMARY

- A vector is with magnitude and direction.

- In space a quantity is specified by a function.

- When the result of multiplication of two vectors is a scalar then it is called scalar product or dot product.

- When product is a vector then it is called vector product.

- Multiplication of three vectors can give scalar $\vec{A} \cdot \left(\vec{B} \times \vec{C} \right)$ or a vector $\vec{A} \times \left(\vec{B} \times \vec{C} \right)$.

- Vector differentiation is done using dal (∇) operator the gradient of a scalar field is $\nabla \phi$, divergence as $\vec{\nabla} \cdot \vec{A}$ and curl by $\vec{\nabla} \times \vec{A}$ and laplacian by $\nabla^2 A$.

- In Cartesion coordinate system

$$\vec{dl} = dx\,\hat{i} + dy\,\hat{j} + dz\,\hat{k}, \quad dV = dx\,dy\,dz$$

$$\text{Gradient } \nabla\phi = \frac{\partial\phi}{\partial x}\,\hat{i} + \frac{\partial\phi}{\partial y}\,\hat{j} + \frac{\partial\phi}{\partial z}\,\hat{k}$$

$$\text{Divergences } \vec{\nabla} \cdot \vec{A} = \frac{\partial A_x}{\partial x} + \frac{\partial A_y}{\partial y} + \frac{\partial A_z}{\partial z}$$

$$\text{Curl } \vec{\nabla} \times \vec{A} = \begin{vmatrix} \hat{i} & \hat{j} & \hat{k} \\ \dfrac{\partial}{\partial x} & \dfrac{\partial}{\partial x} & \dfrac{\partial}{\partial z} \\ A_x & A_y & A_z \end{vmatrix}$$

$$\text{Laplacian } \nabla^2\phi = \frac{\partial^2\phi}{\partial x^2} + \frac{\partial^2\phi}{\partial y^2} + \frac{\partial^2\phi}{\partial z^2}$$

- In cylindrical system

$$\vec{dl} = d\rho\,\hat{\rho} + \rho\,d\phi\,\hat{\phi} + dz\,\hat{z}, \quad dV = \rho\,d\phi\,d\rho\,dz$$

$$\text{gradient } \nabla T = \left[\frac{\partial T}{\partial \rho}\,\hat{\rho} + \frac{1}{\rho}\,\frac{\partial T}{\partial \phi}\,\hat{\phi} + \frac{\partial T}{\partial z}\,\hat{z} \right]$$

$$\text{divergence } \vec{\nabla} \cdot \vec{A} = \frac{1}{\rho}\,\frac{\partial}{\partial \rho}\left(\rho A_\rho \right) + \frac{1}{\rho}\,\frac{\partial A_\phi}{\partial \phi} + \frac{\partial A_z}{\partial z}$$

$$\text{Curl } \vec{\nabla} \times \vec{A} = \frac{1}{\rho} \begin{vmatrix} \hat{\rho} & \hat{\phi} & \hat{z} \\ \dfrac{\partial}{\partial \rho} & \dfrac{\partial}{\partial \phi} & \dfrac{\partial}{\partial z} \\ A_\rho & \rho A_\phi & A_z \end{vmatrix}$$

$$\text{Laplacian } \nabla^2 \phi = \frac{1}{\rho} \frac{\partial}{\partial \rho} \left(\rho \frac{\partial \phi}{\partial \rho} \right) + \frac{1}{\rho^2} \frac{\partial^2 \phi}{\partial \phi^2} + \frac{\partial^2 \phi}{\partial z^2}$$

- In spherical system

$$\vec{dl} = dr\,\hat{r} + rd\theta\,\hat{\theta} + r\sin\theta\,d\phi\,\hat{\phi}$$

$$dV = r^2 \sin\theta\, d\theta\, dr\, d\phi$$

$$\text{gradient } \nabla T = \frac{\partial T}{\partial r}\hat{r} + \frac{1}{r}\frac{\partial T}{\partial \theta}\hat{\theta} + \frac{1}{r\sin\theta}\frac{\partial T}{\partial \phi}\hat{\phi}$$

$$\text{divergence } \vec{\nabla}.\vec{A} = \frac{1}{r^2}\frac{\partial}{\partial r}\left(r^2 A_r\right) + \frac{1}{r\sin\theta}\frac{\partial}{\partial \theta}\left(A_\theta \sin\theta\right) + \frac{1}{r\sin\theta}\frac{\partial A_\phi}{\partial \phi}$$

$$\text{Curl } \vec{\nabla} \times \vec{A} = \frac{1}{r\sin\theta} \begin{vmatrix} \hat{r} & \hat{\theta} & r\sin\theta\,\hat{\phi} \\ \dfrac{\partial}{\partial r} & \dfrac{\partial}{\partial \theta} & \dfrac{\partial}{\partial \phi} \\ A_r & rA_\theta & r\sin\theta\,A_\phi \end{vmatrix}$$

$$\text{Laplacian } \nabla^2 \phi = \frac{1}{\rho}\frac{\partial}{\partial r}\left(r^2 \frac{\partial \phi}{\partial r}\right) + \frac{1}{r^2 \sin\theta}\frac{\partial}{\partial \theta}\left(\sin\theta \frac{\partial \phi}{\partial \theta}\right)$$

$$+ \frac{1}{\partial^2 \sin^2\theta}\frac{\partial^2 \phi}{\partial \phi^2}$$

- Gauss Divergence theorem

$$\oint_s \vec{A}.\overline{ds} = \int_V \left(\vec{\nabla}.\vec{A}\right) dV$$

- Stoke's theorem

$$\oint_L \vec{A}.\overline{dl} = \int_V \left(\vec{\nabla} \times \vec{A}\right).\overline{ds}$$

- A vector field is solenoidal if $\vec{\nabla} \cdot \vec{A} = 0$

  Irrotational or conservative if $\vec{\nabla} \times \vec{A} = 0$

- Triple products

  $$\vec{A} \cdot \left(\vec{B} \times \vec{C}\right) = \vec{B}\left(\vec{C} \times \vec{A}\right) = \vec{C}\left(\vec{A} \times \vec{B}\right)$$

  $$\vec{A} \times \left(\vec{B} \times \vec{C}\right) = \vec{B}\left(\vec{A} \cdot \vec{C}\right) - \vec{C}\left(\vec{A} \cdot \vec{B}\right)$$

- Second derivatives

  $$\vec{\nabla} \cdot \left(\vec{\nabla} \times \vec{A}\right) = 0$$

  $$\vec{\nabla} \times \left(\vec{\nabla} f\right) = 0$$

  $$\vec{\nabla} \times \left(\vec{\nabla} \times \vec{A}\right) = \vec{\nabla}\left(\vec{\nabla} \cdot \vec{A}\right) - \vec{\nabla}^2 \vec{A}$$

EXERCISE

1. Give the basic concepts of transformation of one coordinate system to another. Derive necessary relations for rectangular, cylindrical and spherical systems.   **[RU 2003]**

2. Write short note on "Physical significance of curl, divergence and gradient".

3. State-Gauss divergence theorem. Write its applications, advantages and limitations.
   [RU 2002]

4. State and prove stoke's theorem.   **[RU 2000]**

5. Explain how stoke's theorem enables us to obtain the integral form of ampere circuital law.

6. Explain various types of vector fields.

   (i)   Solenoidal and irrotational fields.

   (ii) Irrotational but not solenoidal fields

   (iii) Solenoidal but not irrotaitonal fields

   (iv) Neither irrotational nor solenoidal fields.

7. For the vectors $\vec{A} = \hat{i} + 3\hat{k}$ and $\vec{B} = 5\hat{i} + 2\hat{j} - 6\hat{k}$ calculate

   (i) $\vec{A} + \vec{B}$ (ii) $\vec{A} - \vec{B}$

   (iii) $\vec{A} \cdot \vec{B}$ (iv) $\vec{A} \times \vec{B}$

   (v)   Angle between $\vec{A}$ and $\vec{B}$

(vi)   A unit vector parallel to $3\vec{A} + \vec{B}$.

(vii) Length of the projection of $\vec{A}$ on $\vec{B}$.

Ans.: $(i)\ 6\hat{i} + 2\hat{j} - 3\hat{k}$ $\qquad$ $(ii) -4\hat{i} - 2\hat{j} + 9\hat{k}$ $\qquad$ $(iii) -13$ $\qquad$ $(iv) -6\hat{i} + 21\hat{j} + 2k$

$(v)\ 60°$ $\qquad\qquad$ $(vi)\ \dfrac{8\hat{i} + 2\hat{j} + 3\hat{k}}{\sqrt{77}}$ $\qquad$ $(vii)\ 1.6m$

8.   Use the differential volume dV to find volume of region.

(i) $0 < x < 1$, $\qquad$ $1 < y < 2$, $\qquad$ $-3 < z < 3$ $\hfill$ [Ans.: **6**]

(ii) $2 < \rho < 5$, $\qquad$ $\dfrac{\pi}{3} < \phi < \pi$, $\qquad$ $-1 < z < 4$ $\hfill$ [Ans.: **110**]

9.   Find area of the region $0 \le \phi \le \beta$ on the spherical shell of radius 'b'. $\hfill$ [Ans. **$2\beta b^2$**]

10. Evaluate the gradient of the following scalar fields

(a) $P = e^{-z} \sin 2x$ $\hfill$ [Ans.: $\nabla P = 2\cos 2x e^{-z}\hat{i} - \sin 2x e^{-z}\hat{k}$]

(b) $q = \rho^2 z \cos \phi$ $\hfill$ $\left[\text{Ans.:} \nabla q = 2\rho z \cos\phi\ \hat{\rho} - \rho^2 z \sin\phi\ \hat{\phi} + \rho^2 \cos\phi \hat{z} \right]$

(c) $s = 20r \sin \theta \cos \phi$ $\hfill$ $\left[\begin{array}{l} \text{Ans.:} \nabla r = 20\sin\theta \cos\phi \hat{r} + 20r\cos\ \phi \\ \quad \cos\theta\ \hat{\theta} - 20r \sin\theta\ \sin\phi\ \hat{\phi} \end{array} \right]$

11. If $f = xy + yz + xz$ then

(i)   Find the magnitude and direction of the maximum rate of change of the function at point (1, 2, 3)

(ii)   Find the rate of change of the function at the same point in the direction of the vector.

$$\left[\begin{array}{l} \text{Ans.:}(i)\ |\nabla f| = \sqrt{14} \\ \qquad (\hat{\nabla}f) = \dfrac{\nabla f}{|\nabla f|} = \dfrac{2\hat{i} + 3\hat{j} + \hat{k}}{\sqrt{14}} \\ (ii)\ df = \nabla f.dl = 11 \end{array} \right]$$

12. If $T = 2x\hat{i} + 3y\ \hat{j} - 4z\hat{k}$ and $V = xyz$ evaluate $\vec{\nabla}.\left(V\vec{T} \right)$. $\hfill$ [Ans.: **2xyz**]

13. If $U = xz - x^2y + y^2z^2$ find div (grad U). $\hfill$ [Ans.: **$2(-y + z^2 + y^2)$**]

14. Given $\vec{D} = 6xyz^2\,\hat{i} + 3x^2z^2\,\hat{j} + 6x^2y\,\hat{k}$ C/m^3 Find the total charge lying within the region bounded by $0 < x < 1$, $1 < y < 2$ and $|z| \le 1$ by separately evaluating each side of divergence theorem. **[Ans.: 6C]**

15. If $\vec{A} = (x+y+1)\hat{i} + \hat{j} - (x+y)\hat{k}$ then prove that $\vec{A} \cdot (\vec{\nabla} \times \vec{A}) = 0$

16. Prove that vector $\vec{A} = (x-yz)\hat{i} + (2y-zx)\hat{j} + (2z-xy)\hat{k}$ is not solenoidal. **[WBUT 2005]**

17. If $\phi = x^2 - y^2 + 2z$ find $\vec{\nabla} \cdot (\vec{\nabla}\phi)$. **[WBUT 2005]**

18. Show that $\vec{B} = 2xyz\,\hat{i} + (x^2z + 2y)\,\hat{j} + x^2y\hat{k}$ is irrotational. **[WBUT 2005]**

19. find a unit vector perpendicular to $x^2 + y^2 - z^2 = 100$ at $(1, 2, 3)$. **[WBUT 2007]**

$$\left[\text{Ans.:}\,\frac{\hat{i} + 2\hat{j} - 3\hat{k}}{\sqrt{14}}\right]$$

20. if $\phi = 3x^2y - y^3z^2$ find $\vec{\nabla}\phi$ at $(1, -2, -1)$. **[WBUT 2004]**

$$\left[\text{Ans.:}\,\vec{\nabla}\phi = -12\hat{i} - 9\hat{j} - 16\hat{k}\right]$$

21. Show that $\vec{F} = (2xy + z^3)\hat{i} + x^2\hat{j} + 3xz^2\hat{k}$ is a conservative force field. Find also the scalar potential. **[WBUT 2006, 2003]** $\left[\text{Ans.:}\,\phi = x^2y + z^3x + \text{constant}\right]$

22. Evaluates $\oint_s \vec{F}.\hat{n}\,ds$ where $\vec{F} = 8x\,z\hat{i} - y^2\hat{j} + yz\hat{k}$ and s is the surface of the cube bounded by x = 0, 1, y = 0, 1, z = 0, 1.

2

ELECTRICITY

(I) Coulomb's law in vector form Electrostatic field and its curt. Gauss's Law in integral form and conversion to differential form. Electrostatic potential and field. Poisson's equation, Laplace equation. (Application to Cartesian, spherical and cylindrically symmetric systems 1-D problem). Electric curved, drift velocity, current density, continuity equation, steady current.

(II) Dielectrics concepts of polarization, the relation $D = \varepsilon_0 E + P$, polarizability, electronic polarization and polarization in monoatomic and polyatomic gases.

Electrostatics

The fundamental physical quality which constitute all electromagnetic fields in charge. It measuring unit is coulomb (c) and characterized by electronic charge where one electron possesses a charge of 1.6×10^{-19} C(i.e. one coulomb of charge is represented by 6×10^{18} electrons). Electronic charge is assumed to be of –ve nature whereas +ve charge is associated with protons in the nucleus.

The complete study of different physical phenomenon between two or more charges when charges are at rest, is known as electrostatics.

When at rest, two charges placed near to each other can experience a force (attractive or repulsive) named as electrostatic forces which can be measured using coulomb's law of electrostatics. The nature of this force is dependent over the types of charges. If the charges are alike, the force will be repulsive a otherwise forces will be attractive.

Electrostatic force-Coulomb's law

Colonial charges tugustion be coulomb in 1785, gave an relation between the magnitude of the force exerted by a point charge q_1 by another charge q in when they are segarated by a distances r in free spaces.

It states that the electrostatic forces between two charges q_1 and q_2 is directly proportional to the product of their charges and inversely proportional to the square of the distance between then and acts along the joining line of two charges.

$$\vec{F} = k \frac{q_1 q_2}{r^2} \hat{r}$$

where k = Coulomb constant

for free space $k = \dfrac{1}{4\pi\varepsilon_0}$

ε_0 = permittivity of free space

$= 8.854 \times 10^{-12}$ Farad/meter. $\quad$ or $\quad \left(\dfrac{(\text{Coulomb})^2}{\text{Newton } (\text{meter})^2} \right)$

so that $k = 9 \times 10^9$ m/f

for any other medium

$$k = \frac{1}{4\pi\,\varepsilon_m}$$

where ε_m – permittivity in the medium

$= \varepsilon_r \varepsilon_0$ where ε_r is relative permittivity of the medium.

Vector form

If two point charges q_1 and q_2 are at a distance $\vec{r}_1$ and $\vec{r}_2$ from the origin, then

$$\vec{r}_{12} = \vec{r}_2 - \vec{r}_1$$

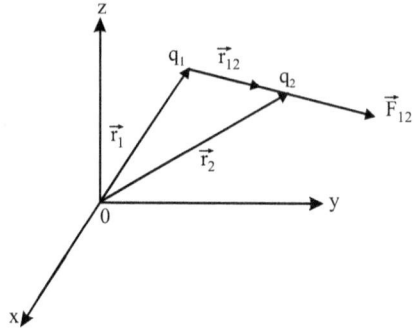

Fig.

so the force $\vec{F}_{12}$ acting on q_2 at $\vec{r}_2$ due to q_1 will be

$$\vec{F}_{12} = \frac{1}{4\pi\varepsilon_0} \frac{q_1 q_2}{r_{12}^2} \hat{r}_{12} = \frac{1}{4\pi\varepsilon_0} \frac{q_1 q_2}{r_{12}^3} \vec{r}_{12} \qquad \text{...(1)}$$

Now $\qquad \vec{r}_{21} = \vec{r}_1 - \vec{r}_2 = -\vec{r}_{12}$

and $\qquad \vec{F}_{21} = \frac{1}{4\pi\varepsilon_0} \frac{q_1 q_2}{r_{21}^2} \hat{r}_{21}$

$$= \frac{1}{4\pi\varepsilon_0} \frac{q_1 q_2}{r_{21}^3} \hat{r}_{21}$$

$$= \frac{1}{4\pi\varepsilon_0} \frac{q_1 q_2}{r_{21}^3} \left(-\vec{r}_{12}\right) \quad \text{as } \vec{r}_{12} = -\vec{r}_{21} \qquad \text{...(2)}$$

and $\qquad \left|\vec{r}_{12}\right| = \left|\vec{r}_{21}\right| = r_{12} = r_{21}$

$$= -\vec{F}_{12}$$

Equation (1) and (2)

CHARGE DISTRIBUTIONS

(1) Linear charge distribution

When a charge Δq is distributed over a conductor of infinitesimally small thickness along length Δl then charge distribution is linear and linear charge density

$$\lambda = \lim_{\Delta l \to 0} \frac{\Delta q}{\Delta l} \text{ C/m}$$

so that the total charge over the entire conductor of length l is

$$q = \int_0^l \lambda \, dl$$

(2) Surface charge Distribution :

When some charge is distributed over a small area then surface charge density will be

$$\sigma = \lim_{\Delta s \to 0} \frac{\Delta q}{\Delta s} \text{ C/m}^2$$

so total charge over the entire surface

$$q = \oint_s \sigma \, ds$$

(3) Volume charge distribution

When charge is distributed over some volume of a conducting medium then volume charge density can be defined as

$$\rho = \lim_{\Delta V \to 0} \frac{\Delta q}{\Delta V} \text{ C/m}^2$$

where ΔV is infinitesimally small volume element so total charge within the volume

$$q = \int_V \rho \, dV$$

Electric Field

When a charge is placed near to another charge, it experiences an electrostatic force according to coulomb's law. The space around any charge upto which any other charge experiences a force, is called electric field or electrostatic field due to that charge.

Mathematically $\qquad \vec{E} = \dfrac{\vec{F}}{q}$ N/C $\qquad\qquad$ ...(1)

If charge q_1 is placed near to charge q_2 then electric field due to q_1 is

$$\vec{E} = \frac{1}{4\pi\varepsilon_0} \frac{q_1 q_2}{r^2} \frac{1}{q_2}$$

$$\vec{E} = \frac{1}{4\pi\varepsilon_0} \frac{q_1}{r^2} \hat{r} \frac{N}{C} \qquad \qquad ...(2)$$

Electric field intensity $\left(\vec{E}\right)$ or the strength of an electric field at a given point is the force exerted or a unit +ve charge placed at that point.

Here it can be summed up that the electrostatic force experienced by a charge q_2 in an electric field $\vec{E}$ is

$$\vec{F} = q_2 \vec{E} \; N \qquad \qquad ...(3)$$

Important Point

(comparison of electrostatic forces with Gravitational force)

1. Electrostatic force can be attractive or repulsive but gravitational force can only be attractive in nature.

2. Charges can be +ve or −ve but mass is always +ve in nature.

3. Coulombs law of electrostatic fine is also a inverse square low similar to Newton's law at Gravitation.

Electric Displacement Vector

The electric displacement vector $\left(\vec{D}\right)$ at any point in a isotropic any linear medium having electrical permittivity ε_m is related with electric fixed intants $\left(\vec{E}\right)$ is given by

$$\vec{D} = \varepsilon_m \vec{E}$$

In free space

$$\vec{D}_0 = \varepsilon_0 \vec{E}$$

SI unit of electric displacement is C/m^2.

Superposition principle

According to this principle when more then one charges interact with unit positive charge, then the net intensity on the text charge is the vector sum of the intensity due to individual charge.

i.e. $\vec{E} = \vec{E}_1 + \vec{E}_2 + ... + \vec{E}_n$

Electric Potential Energy

Electrostatic Potential energy of a system of point charges is equal to the around of work done to bring a text charge from infinity upto a certain point in the electric field of another charge.

Fig.

If two charges are assumed at a distance r_{12}.

Then electrostatic potential energy will be equal to the work done to bring charge q_2 from infinity to point B against the electric field of q_1 as

$$W = \int_{\infty}^{r_{12}} \vec{F}.\,dx$$

$$= -\int_{\infty}^{r_{12}} q_2 \vec{E}.\,dx$$

$$= -q_2 \int_{\infty}^{r_{12}} \frac{q_1}{4\pi\,\varepsilon_0\,x^2}\,dx$$

$$= \frac{1}{4\pi\,\varepsilon_0}\,q_1 q_2 \left[\frac{1}{x}\right]_{\infty}^{r_{12}}$$

$$W = \frac{1}{4\pi\varepsilon_0}\,\frac{q_1 q_2}{r_{12}} \qquad\qquad ...(1)$$

This work done is stored in the system as electrostatic potential energy U.

So $$U = \frac{1}{4\pi\varepsilon_0}\,\frac{q_1 q_2}{r_{12}} \qquad\qquad ...(2)$$

if there is a system formed by more then two charges then the potential energy of the system is

$$U = \frac{1}{4\pi\varepsilon_0}\left[\frac{q_1q_2}{r_{12}} + \frac{q_1q_3}{r_{13}} + \frac{q_2q_3}{r_{23}}....\right] \qquad ...(3)$$

Electric Potential

It is the amount of work done to bring a unit positive charge from infinity to any point in the electric field of another charge.

i.e. It is the electrostatic potential energy per unit text charge

i.e. $$V = \frac{W}{q_2}$$

$$= \frac{1}{4\pi\varepsilon_0}\frac{q_1}{r_{12}}$$

The amount of work done to bring a unit positive text charge from point B to point A against the electrostatic field of charge q_1 is known as electric potential difference.

i.e. $$\int_B^A dV = \int_B^A \vec{E}.\,\vec{dr}$$

so $$V_A - V_B = \frac{q_1}{4\pi\varepsilon_0}\left[\frac{1}{r_A} - \frac{1}{r_B}\right]$$

Electric Flux

The flux of any vector $\left(\vec{A}\right)$ associated with any elementary are $\left(\vec{ds}\right)$ measure the flow of the vector $\left(\vec{A}\right)$ through the elementary area vector is normal direction.

Mathematically

Flux $$\phi = \vec{A}.\,\vec{ds}$$

$$= \vec{A}.\,\hat{n}\,ds$$

where $\hat{n}$ is the unit vector normal to ds.

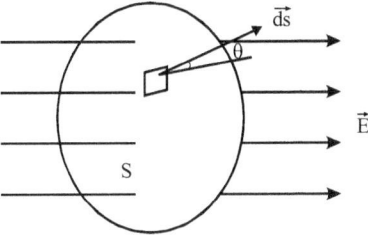

Fig.

When electric field lines moves in a certain direction through any area, then total number of electric lines of force passing normally through the small surface is called electric flux.

So electric flux

$$d\phi = \vec{E} . \vec{ds}$$

total flux through the entire surface

$$\phi = \oint_s \vec{E} . \vec{ds} = \oint_s E \, ds \cos\theta$$

where θ is angle between $\vec{E}$ and $\vec{ds}$.

If the vector is directed outward form the surface, flux is +ve otherwise it is negative.

Conservative force field

When the work done on a particle is independent of the path so that total work done around the closed path is zero, the responsible force is called as conservative force and field developed is called the conservative field.

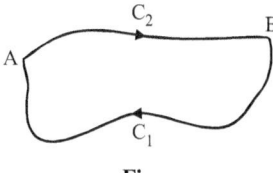

Fig.

Let us assume a particle moving from A to B via C_2 and rheums back through C_1 in a conservational field of force $\vec{F}$.

total work done around the closed path AC_2BC_1 A is zero

$$\oint_{AC_2BC_1A} \vec{F}.\overrightarrow{dr} = \oint_{AC_2B} \vec{F}.\overrightarrow{dr} + \oint_{BC_1A} \vec{F}.\overrightarrow{dr} = 0$$

so

$$\oint_{AC_2B} \vec{F}.\overrightarrow{dr} = \oint_{AC_1B} \vec{F}.\overrightarrow{dr} = 0$$

so work done is independent of path so for conservative for $\oint_C \vec{F}.\overrightarrow{dr} = 0$.

using stoke's theorem

$$\oint_C \vec{F}.\overrightarrow{dr} = \oint_s \left(\vec{\nabla} \times \vec{F}\right).\overrightarrow{ds}$$

So for conservative force

$$\oint_s \left(\vec{\nabla} \times \vec{F}\right).\overrightarrow{ds} = 0$$

or $\qquad\qquad \vec{\nabla} \times \vec{F} = 0 \qquad\qquad$ ...(1)

This is the condition for conservative force field.

for electric potential V (scalar potential)

cure grad V = 0

$$\boxed{\vec{\nabla} \times \vec{\nabla} V = 0} \qquad\qquad \text{...(2)}$$

We can compare equation (1) & (2) and can conclude that from $\vec{F}$ should be a gradient of scalar potential V which is negative as per experimental evidence.

so for conservative field

$$\boxed{\vec{F} = -\vec{\nabla} V} \qquad\qquad \text{...(3)}$$

Curl of electric field $\left(\vec{\nabla} \times \vec{E}\right)$

as electric field $\qquad\qquad \vec{E} = \dfrac{1}{4\pi\varepsilon_0} \dfrac{q}{r^2} \hat{r} \qquad\qquad$ ...(i)

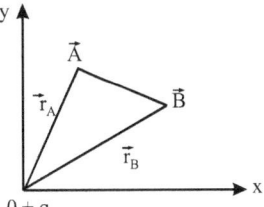

Fig.

line integral of electric field i.e. potential difference between two points A and B will be

$$\int_A^B \vec{E}.\overrightarrow{dr} = \frac{1}{4\pi\varepsilon_0} \int_{r_A}^{r_B} \frac{q}{r^2} dr$$

or
$$\int_{r_A}^{r_B} \vec{E}.\overrightarrow{dr} = \frac{q}{4\pi\varepsilon_0}\left(\frac{1}{r_A} - \frac{1}{r_B}\right) \qquad \qquad ...(ii)$$

if $\overrightarrow{r_A} = \overrightarrow{r_B}$ i.e. A and B are same point.

then
$$\oint_C \vec{E}.\overrightarrow{dr} = 0$$

from stoke's law using $\oint_C \vec{E}.\overrightarrow{dr} = \int_S \left(\vec{\nabla}\times\vec{E}\right).\overrightarrow{ds}$

$$\int_S \left(\vec{\nabla}\times\vec{E}\right).\overrightarrow{ds} = 0$$

or
$$\vec{\nabla}\times\vec{E} = 0$$

Hence electric field is conservative.

Gauss law of electrostatics

Gauss law states that the total electric flux (ϕE) through any closed surface is equal to $\frac{1}{\varepsilon_0}$ times of the total charge enclosed by the surface.

If any arbitrary shaped closed surface S encloses a charge q, then the electric flux through the closed surface is

$$\phi_E = \oint_s \vec{E}.\,\vec{ds} = \frac{q}{\varepsilon_0}$$

Gauss low gives a relation between electric flux and charge.

The imaginary surface, where the magnitude of electric field intensity must be some and normal to the surfaces, is called Gaussian surfaces for any medium of

$$\phi_E = \oint_s \vec{E}.\,\vec{ds} = \frac{q}{\varepsilon}$$

where ε is the permittivity of the medium. as

$$\vec{A} = \varepsilon\vec{E}$$

$$\phi_E = \oint_s \vec{D}.\vec{ds} = q$$

Proof :

Let us consider a point charge +q situated at 0 enclosed by the surface s.

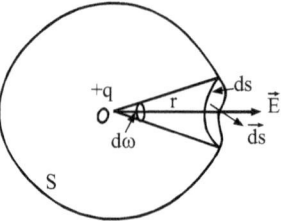

Fig.

if $\vec{E}$ is the electric field intensity at P due to charge +q at O along the direction OP

$$E = \frac{1}{4\pi\varepsilon_0}\frac{q}{r^2} \qquad \qquad ...(i)$$

then outward flux through S

$$\phi_E = \oint_s \vec{E}.\vec{ds} = \oint_s E\,ds\,\cos\theta = \frac{q}{4\pi\varepsilon_0}\oint\frac{ds\,\cos\theta}{r^2} \qquad ...(ii)$$

here $\oint \dfrac{ds \cos \theta}{r^2} = \int dw$ is the solid angle subtended by the entire closed surfaces S at 0 and its value is 4π.

so $\qquad\qquad \phi_E = \oint_s \vec{E} . \overrightarrow{ds} = \dfrac{q}{4\pi\varepsilon_0} \times 4\pi$

$\oint_s \vec{E} . \overrightarrow{ds} = \dfrac{q}{\varepsilon_0}$ which proves the Gauss theorem Gauss theorem can also be stated as the surface integral of the electric field over a closed surface (s) is equal to $\dfrac{1}{\varepsilon_0}$ times of the net charge (q) enclosed by the surfaces.

Differential form of Gauss law

If charge q is distributed over a volume V enclosed by a surface s, then from the definition of volumetric charge distribution, the total charge enclosed by the volume is

$$q = \int_V \rho \, dV \qquad\qquad ...(i)$$

and Guass' law of electrostatics states that

$$\oint_s \vec{E} . \overrightarrow{ds} = \dfrac{q}{\varepsilon_0} \qquad\qquad ...(ii)$$

but from Gauss divergence theorem

$$\oint_s \vec{E} . \overrightarrow{ds} = \int_V \left(\vec{\nabla} . \vec{E} \right) dV \qquad\qquad ...(iii)$$

So using (iii) & (i) in (ii)

$$\int_V \left(\vec{\nabla} . \vec{E} \right) dV = \dfrac{q}{\varepsilon_0} = \dfrac{1}{\varepsilon_0} \int_V \rho \, dV \qquad\qquad ...(iv)$$

so $\qquad\qquad \boxed{\vec{\nabla} . \vec{E} = \dfrac{\rho}{\varepsilon_0}} \qquad\qquad ...(v)$

equation (v) is the differential form of Gaussis law of electrostatics. According to this the divergence of electric field at any point is $\dfrac{1}{\varepsilon_0}$ times the charge density at the point.

as $$\vec{D} = \varepsilon_0 \vec{E}$$

then $$\vec{\nabla} \cdot \left(\varepsilon_0 \vec{E} \right) = \rho$$

so $$\boxed{\vec{\nabla} \cdot \vec{D} = \rho}$$ ...(vi)

This is the differential form of Gauss law in electrostatics in dielectric medium.

Limitation

(a) Here direction of electric field can not be measured as electric flux is a scalar quantity.

(b) This can only be used to find flux through regular/symmetric shaped bodies.

Coulomb's law from Gauss law

When two charges q_1 and q_2 are placed near to each other at some distance r then a gaussian surface of radius r can be drawn with q_1 at its centre A.

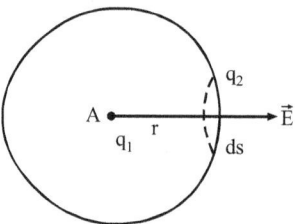

Fig.

then from Gauss theorem

$$\oint_s \vec{E} \cdot \vec{ds} = \frac{q_1}{\varepsilon_0}$$ ...(i)

or $$\oint_s E \, ds \cos 0 = \frac{q_1}{\varepsilon_0}$$

$$E \oint ds = \frac{q_1}{\varepsilon_0} \Rightarrow E.4\pi r^2 = \frac{q_1}{\varepsilon_0}$$

$$\Rightarrow \qquad\qquad E = \frac{q_1}{4\pi\varepsilon_0 r^2} \qquad\qquad\qquad ...(ii)$$

and the force acting on q_2 due to the field

$$F = q_2 E = \frac{q_1 q_2}{4\pi\varepsilon_0 r^2} \qquad\qquad\qquad ...(iii)$$

which is mathematical form of coulomb's law.

Some applications of Gauss' Law of Electrostatics

According to Gauss law $\oint\limits_s \vec{D}.\overrightarrow{ds} = Q$

$\vec{D}.\overrightarrow{ds} = Dds$ when D is perpendicular is surface.

$\vec{D}.\overrightarrow{ds} = 0$ when D is tangential to the surfaces.

(a)   Field due to a point charge

Suppose a point charge is placed at origin then a sphere with centre at origin can be considered as Gaussian surface.

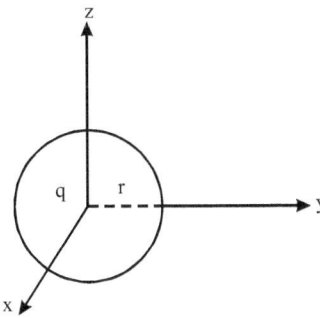

Fig.

so $\qquad \int d\phi = \text{total flux outward}$

$$= \oint_s \vec{D}.\,\overrightarrow{ds} = q$$

$\Rightarrow \qquad \phi = \oint_s \varepsilon_0 \vec{E}.\,\overrightarrow{ds} = q$

$\Rightarrow \qquad \varepsilon_0 \oint E\,ds = q$

$\Rightarrow \qquad \varepsilon_0\,4\pi r^2.E = q$

$\Rightarrow \qquad E = \dfrac{q}{4\pi\varepsilon_0 r^2}$

and $\qquad \boxed{\vec{E} = \dfrac{q}{4\pi\,\varepsilon_0 r^2}\,\hat{r}}$ ...(i)

(b)   Field due to infinite line charge

The Gaussian surface for line charge can be Q cylindrical surface with $\vec{E}$ radially outward over its surface.

The total outward flux is

$$\underset{(1)}{\oint \vec{D}.\overrightarrow{ds}} + \underset{(2)}{\oint \vec{D}.\overrightarrow{ds}} + \underset{(3)}{\oint \vec{D}.\overrightarrow{ds}} = \lambda l = q$$

where λ is linear charge density.

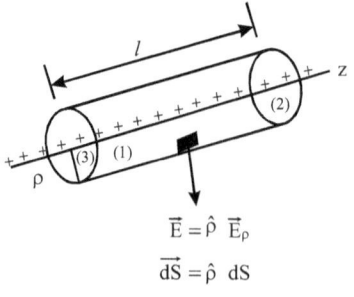

$$\vec{E} = \hat{\rho}\,\vec{E}_\rho$$
$$\overrightarrow{dS} = \hat{\rho}\,dS$$

Fig.

But $\qquad \displaystyle\oint_{(2)} \vec{D}.\overrightarrow{ds} = \oint_{(3)} \vec{D}.\overrightarrow{ds} = 0$

as $\theta = 90°$ between $\vec{D}$ & $\overrightarrow{ds}$.

$\Rightarrow \qquad\qquad \displaystyle\oint_1 \vec{D}.\overrightarrow{ds} = \lambda l$

$\Rightarrow \qquad\qquad \left(\varepsilon_0 \vec{E}_\rho \, \hat{\rho}\right).\hat{\rho}\,\overrightarrow{ds} = \lambda l$

$\Rightarrow \qquad\qquad \varepsilon_0 \, \vec{E}_\rho \, 2\pi \rho l = \lambda l$

$\Rightarrow \qquad\qquad \boxed{\vec{E}_\rho = \dfrac{\lambda}{2\pi\,\varepsilon_0\,\rho}\,\hat{\rho}} \qquad\qquad ...(ii)$

(c)   Field of a charged cloud

Consider a spherical cloud of radius r_0 with uniform charge density ρ c/m³.

(i)   Field outside the cloud i.e. $r > r_0$ –

for s_1

$$\oint_{s_1} \vec{D}.\overrightarrow{ds} = q$$

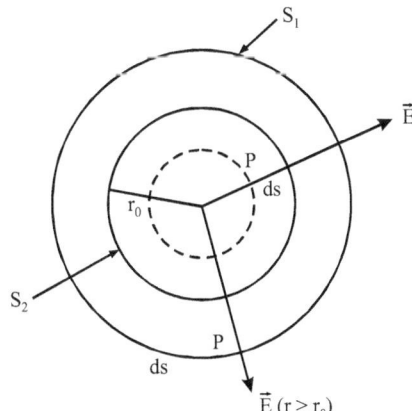

Fig.

or
$$\varepsilon_0 \oint_{s_1} \vec{E}.\,\vec{ds} = \rho \int dv$$

$$= \rho \left(\frac{4}{3} \pi r_0^3 \right)$$

or
$$\varepsilon_0 \, E \, 4\pi r^2 = \rho \frac{4}{3} \pi r_0^3$$

$$E = \frac{\rho \left(4/3 \, \pi r^3 \right)}{4\pi \, \varepsilon_0 r^2} = \frac{\rho r_0^3}{3\varepsilon_0 r^2} \qquad \qquad \ldots(2)$$

or
$$\vec{E} = E \, \hat{n}$$

$$\vec{E} = \frac{\rho \, r_0^3}{3\varepsilon_0 r^2} \hat{n}$$

(ii)   Field inside the cloud i.e. $r < r_0$ –

for closed surface s_2

$$\oint_{s_2} \vec{D}.\vec{ds} = \rho \int_{V_2} dV$$

$$\varepsilon_0 \oint_{s_2} \vec{E}.\vec{ds} = \rho \left[\frac{4}{3} \pi r^3 \right]$$

$$\varepsilon_0 \, E \, 4\pi r^2 = \rho \frac{4}{3} \pi r^3$$

$$E = \frac{\rho r}{3\varepsilon_0}$$

Hence the variation of electric field is as below

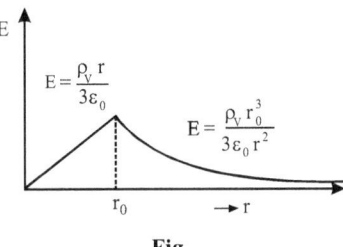

Fig.

(d) Infinite sheet of charge

Suppose there is an infinite sheet of uniform charge σ c/m^2.

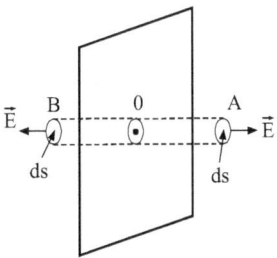

Fig.

the total flux ϕ remains over the (a) two ends of Gaussian cylindrical surface

$$= 2\vec{E} \cdot \vec{ds}$$

$$= 2 \, E \, ds$$

So

$$\oint_s 2E \, ds = \frac{q}{\varepsilon_0} = \frac{\sigma \, ds}{\varepsilon_0} = 2E \, ds$$

or

$$\boxed{E = \frac{\sigma}{2\varepsilon_0}}$$

(e) Field due to a parallel plate capacitor–

A parallel plate capacitor is a combination of two parallel plates separated by some distanced.

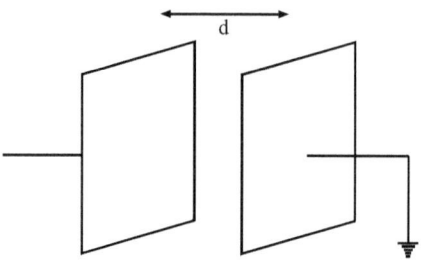

Fig.

So E = 2E₁ where E_1 is electric field due to single plate.

$$= 2.\frac{\sigma}{2\varepsilon_0}$$

⇒

$$\vec{E} = \frac{\sigma}{\varepsilon_0}\,\hat{n}$$

Potential Gradient

Electric field $\vec{E}$ due to +q charge placed at origin 0.

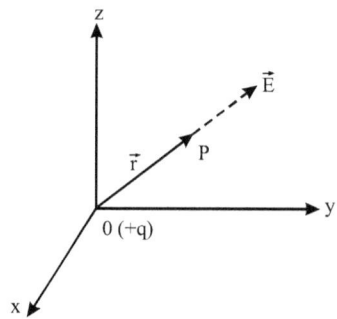

Fig.

$$\vec{E}(r) = \frac{q}{4\pi\,\varepsilon_0 r^2}\,\hat{r} = \frac{q\,\hat{r}}{4\pi\,\varepsilon_0 r^2} \qquad \text{...(i)}$$

if $\vec{r}$ is the position vector of point (x, y, z)

then
$$r = \sqrt{x^2 + y^2 + z^2}$$

and
$$\vec{\nabla}\left(\frac{1}{r}\right) = \vec{\nabla}\left(\frac{1}{\sqrt{x^2 + y^2 + z^2}}\right)$$

$$= \vec{\nabla}\left(\left(x^2 + y^2 + z^2\right)^{-1/2}\right)$$

$$= \frac{-\left(x\hat{i} + y\hat{j} + z\hat{k}\right)}{\left(x^2 + y^2 + z^2\right)^{3/2}}$$

$$= -\frac{\vec{r}}{r^3} \qquad \qquad \text{...(ii)}$$

so using (ii) in (i)

$$\vec{E}(r) = -\frac{q}{4\pi\varepsilon_0}\,\vec{\nabla}\left(\frac{1}{r}\right)$$

$$= -\vec{\nabla}\left(\frac{q}{4\pi\varepsilon_0 r}\right)$$

$$= -\vec{\nabla}V$$

hence
$$\boxed{\vec{E}(1) = -\vec{\nabla}V}$$

(a)   −ve sign indicates that the direction of $\vec{E}$ is along the direction of decreasing V.

(b)   When $\vec{E} = 0$ then $\vec{\nabla}V = 0 \Rightarrow V =$ constant. so theotric potential will have constant value not compulsorily zero.

Poisson's and Laplace Equation

According to the differential form of Gauss law

$$\vec{\nabla}.\vec{E} = \frac{\rho}{\varepsilon_0} \qquad \qquad \text{...(i)}$$

and $$\vec{E} = -\vec{\nabla} V$$ ...(ii)

using (ii) in (i)

$$\vec{\nabla}.\left(-\vec{\nabla}V\right) = \frac{\rho}{\varepsilon_0}$$

or $$\boxed{\nabla^2 V = -\frac{\rho}{\varepsilon_0}}$$ ...(iii)

or $$\text{div grad } V = -\frac{\rho}{\varepsilon_0}$$

This equation (iii) is known as Poisson equation. Any static electric field should satisfy this equation which gives a relation between potential and charge density at any point in an electric field in space.

For a charge free region $\rho = 0$

$$\boxed{\nabla^2 V = 0}$$ ...(iv)

equation (iv) is known as Laplace's equation.

The solution of this equation gives the potential in charge free space.

as $\nabla^2 V = 0 \Rightarrow V = $ constant, so potential remains constants at all points.

The Laplace equation in three coordinates are

(i) Cartessian coordinates

$$\nabla^2 V = \frac{\partial^2 V}{\partial x^2} + \frac{\partial^2 V}{\partial y^2} + \frac{\partial^2 V}{\partial z^2} = 0$$ ...(v)

(ii) Cylindrical coordinates

$$\nabla^2 V = \frac{1}{\rho}\frac{\partial}{\partial \rho}\left(\rho\frac{\partial V}{\partial \rho}\right) + \frac{1}{\rho^2}\frac{\partial^2 V}{\partial \phi^2} + \frac{\partial^2 V}{\partial z^2} = 0$$ ...(vi)

(iii) Spherical coordinates

$$\nabla^2 V = \frac{1}{r^2} \frac{\partial}{\partial r}\left(r^2 \frac{\partial V}{\partial r}\right) + \frac{1}{r^2 \sin \theta} \frac{\partial}{\partial \theta}\left(\sin \theta \frac{\partial V}{\partial \theta}\right)$$

$$+\frac{1}{r^2 \sin^2 \theta} \frac{\partial^2 V}{\partial \phi^2} = 0 \qquad \qquad ...(viii)$$

Applications of Laplace equation in Electrostatics–

(i)   A parallel plate capacitor–

Consider a parallel plate capacitor having its plates at $z = 0$ and $z = d$ with upper plates potential at V_1 and lower feet grounded as shown

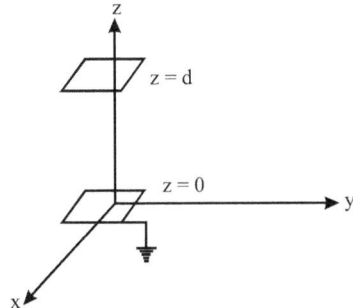

Fig.

here $X - Y$ component of potential are zero so

$$\frac{\partial^2 V}{\partial x^2} = \frac{\partial^2 V}{\partial y^2} = 0$$

hence from Laplace equation in Cartesian coordinates

$$\frac{\partial^2 V}{\partial z^2} = 0$$

Integrating $\dfrac{\partial V}{\partial z} = C$ and again

$$V = Cz + D \qquad \qquad ...(i)$$

Using boundary conditions

at $\quad$ z = 0, $\quad$ V = 0, $\quad$ so D = 0

and at $\quad$ z = d, $\quad$ V = V$_1$ $\quad$ so C = $\dfrac{V_1}{d}$ $\qquad$ ...(ii)

so using (ii) in (i)

$$V = \frac{V_1 z}{d}$$ $\qquad$ ...(iii)

(ii) A spherical capacitor

Consider any point P inside a sphere at a distance r from the centre of the sphere. The variation of potential V is considered only along radial direction.

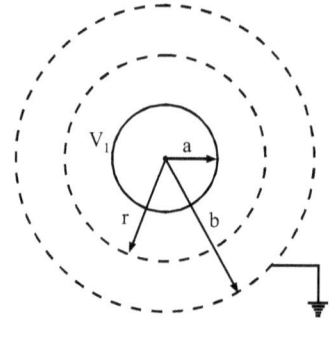

Fig.

So Laplace equation in spherical coordinates as

$$\nabla^2 \equiv \frac{1}{r^2} \frac{\partial}{\partial r} \left(r^2 \frac{\partial V}{\partial r} \right) = 0$$

or $\qquad \dfrac{\partial}{\partial r} \left(r^2 \dfrac{\partial V}{\partial r} \right) = 0$ $\qquad$ ...(i)

Integrating

$$r^2 \frac{\partial V}{\partial r} = \text{constants} = C$$

$\Rightarrow$ $$\frac{\partial V}{\partial r} = \frac{C}{r^2}$$

Integrating

$$V = -\frac{C}{r} + D \qquad \qquad ...(ii)$$

applying boundary conditions

at $r = b$, $V = 0$

$$0 = -\frac{C}{b} + D$$

or $$D = \frac{C}{b}$$

so $$V = -\frac{C}{r} + \frac{C}{b} = C\left[\frac{1}{b} - \frac{1}{r}\right] \qquad \qquad ...(iii)$$

at $r = a$, $V = V_1$

so $$V_1 = C\left[\frac{1}{b} - \frac{1}{a}\right]$$

or $$C = \frac{V_1}{\left(\frac{1}{b} - \frac{1}{a}\right)} \qquad \qquad ...(iv)$$

Using (iii) & (iv) in (ii)

$$V = \frac{V_1}{\left(\frac{1}{b} - \frac{1}{a}\right)}\left(\frac{1}{b} - \frac{1}{r}\right)$$

or $$V = \frac{V_1}{k}\left(\frac{1}{b} - \frac{1}{r}\right) \quad \text{where} \quad k = \frac{1}{b} - \frac{1}{a} = \frac{a-b}{ab} \qquad \qquad ...(v)$$

Which is the expression of potential inside a spherical capacitor.

(iii) Cylindrical capacitor

Consider any point P inside a cylinder at distance r with variation of potential only along the radial direction only. So Laplace equation in cylindrical coordinates reduces to

$$\nabla^2 V \equiv \frac{1}{r}\frac{\partial}{\partial r}\left(r\frac{\partial V}{\partial r}\right) = 0$$

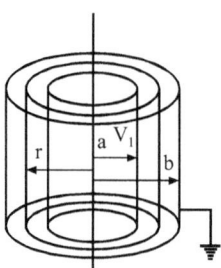

Fig.

or

$$\frac{\partial}{\partial r}\left(r\frac{\partial V}{\partial r}\right) = 0 \qquad \text{...(i)}$$

Integrating (i) ⇒

$$r\frac{\partial V}{\partial r} = \text{constant} = C$$

or

$$\frac{\partial V}{\partial r} = \frac{C}{r}$$

Integrating again

$$V = C \ln r + D \qquad \text{...(ii)}$$

applying boundary conditions

at r = b, v = 0

$$0 = C \ln r + D$$

or

$$D = -C \ln b \qquad \text{...(iii)}$$

Using (iii) in (ii)

$$V = C \ln r - C \ln b$$

$$= C \ln \frac{r}{b} \qquad \ldots(iv)$$

when $r = a$, $V = V_1$

$$V_1 = C \ln \frac{a}{b}$$

or
$$C = \frac{V_1}{\ln \frac{a}{b}} \qquad \ldots(v)$$

So from (ii)

$$V = \frac{V_1 \ln \frac{r}{b}}{\ln \frac{a}{b}} \qquad \ldots(vi)$$

This gives the potential at a point inside the cylindrical capacitor with $a < r < b$.

Electric current

There are large number of free electrons available due to overlapping of conduction bend and valence band in metals. For metal energy band gap $\Delta E_g = 0$ so free electrons move inside the metal in all directions and the net rate of movement is zero. Hence in absence of any external electric field there is no net banofer of charge. but when an external electric field in applied then due to electrostatic force these free electrons get accelerated and move towards opposite to the electric field. The rate of flow of electric charge through any cross section of the conductor is known as electric current (I).

i.e.
$$I = \frac{dq}{dt} \qquad \ldots(i)$$

If this current (I) is uniformly distributed arose conductor of cross sectional area A then the current density at all points in this cross section is defined as the current flowing per unit area of cross section of the conductor.

$$J = \frac{I}{A} \quad \text{which is a vector quantity}$$

$$I = \oint_s \vec{J}.\, \vec{ds}$$

It represents the current flowing through the entire surface s.

Equation of Continuity

If ρ is the charge density, then in a small volume element dV, then total charge in this volume

$$q = \int_V \rho\, dV \qquad \qquad ....(i)$$

Current I si the rate of decrease of charge from the volume of the conductor

$$I = -\frac{dq}{dt} \qquad \qquad ...(ii)$$

so

$$I = -\frac{d}{dt} \int_V \rho\, dV = -\int_V \frac{\partial \rho}{\partial t} dV \qquad \qquad ...(iii)$$

If $\vec{j}$ is the currents density then current is

$$I = \oint_s \vec{J}.\, \vec{ds} \qquad \qquad ...(iv)$$

so from (iii) & (iv)

$$\oint_s \vec{J}.\, \vec{ds} = -\int_V \frac{\partial \rho}{\partial t} dV \qquad \qquad ...(v)$$

Form Gauss's Divergence theorem

$$\oint_s \vec{J}.\, \vec{ds} = \int_V \left(\vec{\nabla}.\vec{J}\right) dV \qquad \qquad ...(vi)$$

So from (v) using (vi)

$$\int_V \left(\vec{\nabla}.\vec{J}\right) dV = -\int_V \frac{\partial \rho}{\partial t} dV$$

or $\qquad \int\limits_V \left(\vec{\nabla}.\vec{J} + \dfrac{\partial \rho}{\partial t} \right) dV = 0$ ...(vii)

or $\qquad \boxed{\vec{\nabla}.\vec{J} + \dfrac{\partial \rho}{\partial t} = 0}$ ...(viii)

This represents the equation of continuity which provides a relation between charge density and current density.

Equation (viii) implies that decrease of charge inside a volume of a conductor with time is equal to the flow of charge from that surface enclosing the volume, which satisfies the conservation of charge.

Drift Velocity

Let us consider a conductor of length l and cross section area A.

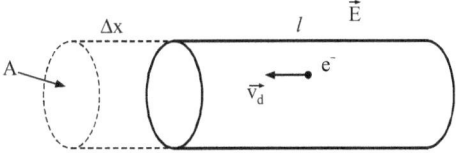

Fig.

In a conductor the electrons are in random motion in all directions due to which the net effect is zero. But when an electric field is applied across the conductor free electrons are drifted towards the electric field with a certain velocity. During this motion electron collides with other particle and again comes to just and then under electrostatic force again starts to move in any random direction. The line elapsed between two successive collisions is known as relaxation time. Overall electron drifts towards a certain direction of electric field.

The drift velocity is given by

$$v_d = \left(u_1 + u_2 + ... \right) + \dfrac{a\left(\tau_1 + \tau_2 + \tau_3 + ... \right)}{n}$$

where u_1, u_2, u_3 .... are initial velocities of electron after collision, a is the acceleration generated due to electrostatic force and τ_1, τ_2, τ_3 are relaxation times during two successive collisions.

$$u_1 = u_2 = u_3 ... = 0$$

so
$$v_d = \frac{a(\tau_1 + \tau_2 + \tau_3 + ...)}{n}$$

$$= a\tau \text{ where } \tau \text{ is the mean relaxation time.} \qquad ...(i)$$

the electrostatic force

$$\vec{F} = e\vec{E} \qquad ...(ii)$$

so
$$eE = a\tau \Rightarrow a = \frac{eE}{m} \qquad ...(iii)$$

using (iii) in (i)

$$v_d = \frac{eE}{m}\tau \qquad ...(iv)$$

due to this drifting of electrons a current is setup across the conductor known as steady current if there are n free electrons per unit volume then the total no. of free electrons passing through cross station area A in time Δt.

$$N = n A x$$

$$= n A v_d \Delta t \qquad ...(v)$$

total charge passing through

$$\Delta q = ne A v_d \Delta t \qquad ...(vi)$$

hence the rate of flow of charge i.e. steady current

$$I = \frac{\Delta q}{\Delta t}$$

$$= \frac{ne A v_d \Delta t}{\Delta t}$$

$$I = ne A v_d \qquad ...(vii)$$

using
$$v_d = \frac{eE}{m}\tau \text{ in (vii)}$$

$$I = \frac{ne^2 AE}{m} \tau \qquad \text{...(viii)}$$

According to Ohm's claw the potential difference V and resistance R is heated with I as

$$V = IR$$

and potential difference $\quad V = El$

$$\Rightarrow \qquad E = \frac{V}{l} \qquad \text{...(ix)}$$

so

$$I = \frac{ne^2 A V}{m\, l} \tau$$

or

$$\frac{V}{I} = \frac{m\, l}{n\, e^2\, \tau\, A} = R$$

so

$$R = \frac{m}{n\, e^2 \tau} \frac{l}{A} = \rho\, \frac{l}{A} \qquad \text{...(x)}$$

where $\qquad \rho = \dfrac{m}{n e^2 \tau}$ is the resistivity of the conductor.

$$\text{Current density } \vec{j} = \frac{I}{A} = \frac{ne^2 \vec{E}}{m} \tau$$

or

$$\boxed{\vec{J} = \sigma \vec{E}} \qquad \text{...(xl)}$$

where $\qquad \sigma = \dfrac{n e^2 \vec{E}}{m}$ is conductivity of the conductor.

Average drift velocity per unit electric field is called as mobility

as $\qquad v_d \propto E$

so $\qquad v_d = \mu_e E$

or $\qquad \mu_e = \dfrac{v_d}{E}$

Where μ_e is the mobility of electron.

DIELECTRICS

As we know that conductors have a very large number of free electrons so that electrons can move in the material without any problem of electrostatic force as it is negligible and there is no requirement to apply external electric field to move there free electrons.

On the other hand insulators do not have any free electrons and conduction is not possible even after the application of external electric field, but there are certain materials which are insulators in all conditions but when some external field is applied on them they get polarized and show conducting properties and an electric field is setup inside the material so as to oppose the external electric field. There are known as dielectric materials. So, dielectric materials are those which are electrical insulators and an electrical field can be setup in that without much dissipation of power.

This property of dielectric materials is used to construct radio frequency transmission lines and to develop capacitors especially at radio frequencies.

Hence dielectric material in defined as a special material which remains insulator under almost conditions and can be polarized and electrostatic field can be setup in it for a long time.

e.g. parafinware, air, paper, glass, mics etc.

The resistivity of the dielectric material is in the range of 10^6 Ω-m to 10^{16} Ω-m.

Effect of a dielectric on the behaviour of a capacitor

Let us consider a parallel plate capacitor with vacuum between the plates with capacitance C_0 given by

$$C_0 = \frac{Q_0}{V} = \frac{\varepsilon_0 A}{d}$$

where A is the plate area and ε_0 is the electrical permittivity of vacuum.

Whenever a dielectric slab is inserted in between the plates, keeping V same then charge increases from Q_0 to Q on the plates.

so capacitance increases by the ratio Q to Q_0.

then
$$\varepsilon_r = \frac{Q}{Q_0} = \frac{C}{C_0}$$

where ε_r is the relative permittivity.

$$\varepsilon_r = \frac{\varepsilon_m}{\varepsilon_0} \quad \text{ratio of permittivity in the medium to free space.}$$

$$\varepsilon_0 = \frac{1}{c^2 \mu_0} = 8.854 \times 10^{-12} \text{ F/m}$$

where C is speed of light and μ_0 is permeability of free space.

This relative permittivity in defined as the dielectric constant i.e. $k = \frac{\varepsilon}{\varepsilon_0}$.

Hence the dielectric constant of a material is the ratio of the capacitance of a given capacitor field completely with that material to the capacitance in vacuum or in other words, the ratio of permittivity of medium to that of the vacuum is known as dielectric constant.

If is found to be independent of the shape and dimension of the capacitor.

Dielectric constants of some materials (Table)

S. No.	Material	Dielectric constant k	Temp (F)
1.	Vacuum	1	-
2.	Air dry	1	68
3.	Polyethylene	2.25	68
4.	Paper	3	68
5.	Teflon	2	75
6.	Paraffin	2.3	75
7.	Petroleum sit	2	75
8.	Water	80	68
9.	Mica	7	75
10.	Glycerin	47	68
11.	Rubber	7	68
12.	Methyl Alcohol	30	68
13.	Barium titanate	1200	68

Dipole movement and polarization

An electrical dipole is an arrangement where two equal and opposite charges are held at very small distance to each other.

When any dielectric substances is kept inside any electric field then the effect of this field induces electrical dipoles in the material and try to align them into field direction. In addition to the generation of new electric dipoles, external field also try to align already existed dipoles and we have the combined effect. This total effect of an external electric field on any dielectric material is called polarization of the dielectric substance.

Any electrical dipole is characterized by its dipole moment which is the product of the magnitude of the charge and the separation between the centre of masses of +ve and −ve charges hence dipole moment.

$$\vec{p} = q\,\vec{r} \qquad \qquad ...(1)$$

It is directed from negative to positive charge and hence it is a vector quantity. It acid is debye (D).

where, $\qquad\qquad$ debye $= 10^{-18}$ state c. cm.

$$\approx 3.33 \times 10^{-30}$$

Centre of mass of
+ve and −ve
charges coincides

separation of centre
of masses of charges
due to electric field.

Newly developed
electric dipole

Fig.

The electric dipole moment per unit volume is called polarization or polarization diversity $(\vec{p})$. It is always directed from negative charge to positive charge.

If there are N atoms per unit volume than

$$\vec{p} = N\vec{p} \qquad \qquad ...(ii)$$

where $\vec{p}$ is the electric dipole moment of individual atom.

The dielectric molecules become polarized when it is placed under some external electric field. From The polarization vector $(\vec{p})$ is proportional to electric field experienced by the dielectric modulus so

$$\vec{p} \propto \vec{E}$$

$$\vec{p} = X \, \varepsilon_0 \, \vec{E} \qquad \qquad ...(iii)$$

Where X is called the electrical susceptibility of the dielectric material. It is equal to the ratio of polarization per unit volume to electrical intensity in the dielectric.

Again, the net induced dipole moment $(\vec{p})$ of an atom of dielectric substance placed in an electric field is proportional to the applied field $(\vec{E})$ with its direction parallel to the field so that

$$\vec{p} \propto \vec{E} \text{ or } \vec{p} = \alpha \, \vec{E} \qquad \qquad ...(iv)$$

where α is called atomic polarizability $\left(\alpha = \dfrac{\vec{p}}{\vec{E}} \right)$.

Hence atomic polarizability is equal to the induced dipole moment for an atom when electric field of unit strength is applied on it.

$$\alpha = \frac{p}{E} = \frac{c.m}{Vm^{-1}} = CV^{-1}m^2 = Fm^2$$

so form (ii)

$$\vec{p} = n \propto \vec{E} \qquad \qquad ...(v)$$

As we know that polarization density

$$p = \frac{\text{Total dipole moment}}{\text{Volume of dielectric step}}$$

$$= \frac{qd}{sd} = \frac{q}{s} = \sigma_p$$

so $\qquad \qquad \qquad p = \sigma_p \qquad \qquad ...(vi)$

When dielectric slab is placed inside a parallel plate capacitor, the effective electric field is reduced by

$$E = \frac{\sigma - \sigma_p}{\varepsilon_0} = \frac{\sigma}{\varepsilon_0} - \frac{\sigma_p}{\varepsilon_0}$$

$\Rightarrow$ $\qquad E = E_0 - \dfrac{\sigma_p}{\varepsilon_0}$ $\qquad \left(\text{or } E = E_0 - \dfrac{P}{\varepsilon_0} \right)$ ...(vii)

$\Rightarrow$ $\qquad \varepsilon_0 E = \varepsilon_0 E_0 - \sigma_p$

$\Rightarrow$ $\qquad \varepsilon_0 E_0 = \varepsilon_0 E - \sigma_p$

using (vi)

$$\varepsilon_0 E_0 = \varepsilon_0 E - p$$

But $\varepsilon_0 E_0$ is known as the electrical displacement vector D so

$$\boxed{\vec{D} = \varepsilon_0 \vec{E} + \vec{P}}$$ ...(viii)

As $\qquad \chi = \dfrac{\vec{p}}{\varepsilon_0 \vec{E}}$

$\Rightarrow$ $\qquad \vec{p} = \chi \, \varepsilon_0 \vec{E}$

placing in above (vii)

$$E = E_0 = \frac{\chi \, \varepsilon_0 E}{\varepsilon_0} = E_0 - \chi E$$

or $\qquad \dfrac{E_0}{E} = 1 + \chi$

$\Rightarrow$ $\qquad \boxed{k = 1 + \chi}$ ...(ix)

where $\qquad k = \dfrac{E_0}{E}$ is the dielectric constant of the dielectric material.

TYPES OF DIELECTRICS

A molecule is assumed to be neutral where the algebraic sum of all the charges is zero.

Any material can have a large number of molecules and each molecule consist of a nucleus and electrons. Depending on the separation between the centre of gravity of positive and negative charges the material can be classified as polar or non polar dielectric material.

(i)   Non-polar dielctric

A molecule in which the centre of gravity of positive and negative charges coincides, due to which the molecule does not possess any permanent dipole moment in the absence of external electric field. There are called non-polar molecule and the material as non-polar dielectric material. these generally have symmetrical arrangement.

Monoatomic molecules like He, Ne, Ar, Xe or molecules having two identical atoms like H_2, N_2, O_2, cl_2 etc. are non-polar. CO_2 due to its linearly symmetric structure is also non-polar.

(ii)   Polar dielectrics

Here the centre of gravity of positive charge is finitely separated from that of negative charge, resulting in an electric dipole with some dipole moment. which is permanent.

Materials made up of there molecules are called polar dielectrics.

Polyatomic molecules like N_2O, H_2O, HCl, NH_3 etc. are polar.

Types of polarization

Some of the important polarizations are as under–

(i)   Electronic polarization

Here when the external field is applied, the electron clouds of atom are displaced with respect to the heavy nuclei within the dimensions of atom. this is called electronic polarization. It does not depend upon temperature.

$$\vec{p}_c = N\,\alpha_e \vec{E}$$

(ii)   Ionic polarization

It occurs only in some ionic crystals. In the presence external electric field the positive and negative ions are displaced upto the point where ionic bonding force stop this displacement. Hence dipoles gets induced. There also do not depend upon temperature.

(iii)   Orientational polarization

It applies only in polar dielectric materials. Generally, in absence of external electric field electric dipoles are so oriented randomly that their net effect becomes zero but in presence of electric field, these dipole try to rotate and align in the direction of electric field. This is known as orientation polarization which is dependent over temperature also.

The total polarization in the sum of all there effects as $\vec{P} = \vec{P}_c + \vec{P}_s + \vec{P}_0$

Relation between electronic polarizability and atomic radius

Let us consider an atom with atomic number z and radius r.

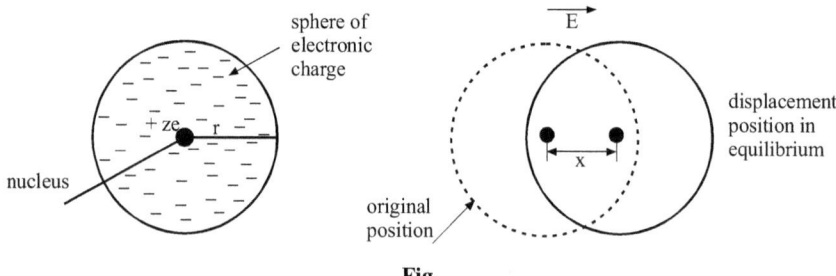

Fig.

have charge density

$$\rho = \frac{-ze}{\frac{4}{3}\pi r^3}$$

when this atom is placed inside any electric field $\vec{E}$, there is a displacement of x between nucleus and electron cloud. The responsible force

$$F = (ze)\ E \qquad \qquad \text{...(i)}$$

$$= \frac{(ze)\left(\frac{4}{3}\pi x^3 \cdot \rho\right)}{4\pi\varepsilon_0 x^2}$$

$$= -\frac{z^2 e^2 x}{4\pi\ \varepsilon_0\ r^3} \qquad \qquad \text{...(ii)}$$

at equilibrium the nucleus is balanced so total effect of force must be zero.

$$zeE = +\frac{z^2 e^2 x}{4\pi\ \varepsilon_0\ r^3}$$

or $$x = \frac{4\pi\,\varepsilon_0\,r^3}{ze} \qquad \qquad ...(iii)$$

which shows displacement x α applied   field E so the induced electronic dipole moment

$$p = (ze)x$$

or $$\vec{p} = 4\pi\,\varepsilon_0\,r^3\,\vec{E} \qquad \qquad ...(iv)$$

Hence induced electronic dipole moment is proportional to the applied field strength.

Now electronic polarizability

$$\alpha_e = \frac{\vec{p}}{\vec{E}}$$

$$\alpha_e = 4\pi\,\varepsilon_0 r^3 \qquad \qquad ...(v)$$

if there are n atoms per cubic meter, the total electronic polarization

$$\vec{p} = n\vec{p}$$

$$= n\,\alpha_e\,\vec{E}$$

$$\vec{p} = n\left(4\pi\varepsilon_0 r^3\right)\vec{E} \qquad \qquad ...(vi)$$

Gauss law in dielectric materials

Gauss law status that the total electric flux ϕ passing through a closed surface is equal to the total charge enclosed by that surface.

i.e. $$\phi = \oint_s \vec{E}.\overrightarrow{ds} = \frac{q}{\varepsilon_0}$$

where q is the total charge enclosed

so $$\oint_s \vec{E}.\overrightarrow{ds} = q$$

or $$\oint_s \vec{D}.\overrightarrow{ds} = q$$

SOLVED EXAMPLES

Q.1: Find the total charge contained within a sphere of radius 'r' and volume charge density proportional to radius.

Sol.: As volume charge density α radius

so
$$\rho \propto r$$

$$\rho = Cr$$

C is constant of proportionality we know that differential volume element in spherical coordinate system

$$dV = r^2 \sin \theta\, dr\, d\theta\, d\phi$$

hence small charge element

$$d\theta = \rho \times dv$$

$$= Cr \times r^2 \sin \theta\, dr\, d\theta\, d\phi$$

$$= Cr^3 \sin \theta\, dr\, d\theta\, d\phi$$

hence total charge

$$Q = \int_0^r \int_{\theta=0}^{\pi} \int_{\phi=0}^{2\pi} C\, r^3 \sin \theta\, d\theta\, dr\, d\phi$$

$$= C \int_0^r r^3 dr \int_0^{\pi} \sin \theta\, d\theta \int_0^{2\pi} d\phi$$

$$= C \cdot \frac{r^4}{4} \cdot \cos \theta \Big|_0^{\pi}\ 2\pi$$

$$\theta = \pi\, C r^4$$

Q.2 : Two unknown point charges are placed at a distance r apart. Find the ratio of the charges at a point on the line joining them where electric at a point on the line joining them where electric field strength becomes zero. Also comment about their natures.

Sol.: If q_1 and q_2 are the two charges.

then

Fig.

so
$$\frac{q_1}{4\pi\,\varepsilon_0 x^2} = \frac{q_2}{4\pi\,\varepsilon_0\left(r-x\right)^2}$$

or
$$\frac{q_2}{q_1} = \left(\frac{r-x}{x}\right)^2$$

or
$$q_2 : q_1 = \left(r-x\right)^2 : x^2$$

as $\left(\dfrac{r-x}{x}\right)^2$ will always be +ve so both the charges will either +ve or –ve.

Q.3: Point charges 5nC and 2nC are located at (2, 0, 4) and (–3, 0, 5) respectively. Find

(a)   Force on 1 nC point charge located at (1, –3, 7).

(b)   Electric field at (1, –3, 7).

[RU 2006]

Sol.:

Let $q_1 = 5$ nC   at (2, 0, 4)

so $\vec{r_1} = 2\hat{i} + 4\hat{k}$

and $q_2 = 2$nC at$\left(-3,0,5\right)$

so $\vec{r_2} = -3\hat{i} + 5\hat{k}$

and $q = 1$nC   at (1, –3, 7)

so $\vec{r} = \hat{i} - 3\hat{j} + 7\hat{k}$

so that $\vec{r} - \vec{r_1} = \hat{i} - 3\hat{j} + 3\hat{k}$

and $$\vec{r} - \vec{r}_2 = 4\hat{i} - 3\hat{j} + 2\hat{k}$$

(i) According to superposition principle

$$\vec{F} = \sum_{i=1}^{2} k \frac{qQ_i}{|r - r_i|^3} (\vec{r} - \vec{r}_i)$$

$$= 9 \times 10^9 \times 1 \times 10^{-9} \left[\frac{5 \times 10^{-9} \times (\hat{i} - 3\hat{j} + 3\hat{k})}{\left[\sqrt{1^2 + (-3)^2 + 3^2} \right]^3} + \frac{2 \times 10^{-9} \times (4\hat{i} - 3\hat{j} + 2\hat{k})}{\left[\sqrt{4^2 + (-3)^2 + 2^2} \right]^3} \right]$$

$$= 9 \times 10^9 \left[\frac{5}{19^{3/2}} (-\hat{i} - 3\hat{j} + 3\hat{k}) + \frac{2}{29^{3/2}} (4\hat{i} - 3\hat{j} + 2\hat{k}) \right]$$

$$= 9 \times 10^9 \left[-0.008\hat{i} - 0.2184\hat{j} + 0.2056\hat{k} \right]$$

$$= \left[-0.072\hat{i} - 1.965\,\hat{j} + 1.85\,\hat{k} \right] \times 10^{-9} \text{ Newtons}$$

(ii) Electric field at that point

$$\vec{E} = \frac{\vec{F}}{q} = \frac{\left[-0.075\hat{i} - 1.965\hat{j} + 1.85\hat{k} \right] \times 10^{-9}}{1 \times 10^{-9}}$$

$$= \left[-0.072\hat{i} - 1.965\,\hat{j} + 1.85\,\hat{k} \right] \text{V/m}$$

Q.4 : If the electric field on a region is $\vec{E} = 4\hat{i} + 6\hat{j} + 7\hat{k}$, find the electric flux through the surface area of 75 square units in XY plane.

Sol.: Surface are in XY plane means that area vector is along z-direction.

So $$\vec{A} = 75\hat{k}$$

here $$\vec{E} = 4\hat{i} + 6\hat{j} + 7\hat{k}$$

so electric flux $\qquad \phi_E = \displaystyle\oint_s \vec{E} \cdot \vec{ds}$

$$= \vec{E} \cdot \vec{A}$$

$$= \left(4\hat{i} + 6\hat{j} + 7\hat{k}\right) \cdot \left(75\hat{k}\right)$$

$$= 525 \ \text{Nm}^2\text{C}^{-1}$$

Q.5 : If 2000 flux lines enter through a given volume of space and 4000 lines diverge from it, calculate the total charge within the volume.

Sol.: Let $\phi_1 = 2000$ Vm and $\phi_2 = 4000$ Vm.

According to Gauss's law of electrostatics

Total flux $\qquad \phi = \dfrac{q}{\varepsilon_0} = \phi_2 - \phi_1$

$$q = \varepsilon_0 \left(\phi_2 - \phi_1\right)$$

$$= 8.85 \times 10^{-12} \times 2000$$

$$= 1.77 \times 10^{-8} \text{C} \cdot$$

Q.6 A hollow metallic sphere of radius 0.1 m has 1×10^{-8}C of charge distributed uniformly over it. Calculate electric field intensity (i) on the surface of the sphere (ii) at 8 cm away from the centre and (iii) at point 1 m away from the centre.

Sol.: (i) Electric field intensity on the surface

$$E = \dfrac{1}{4\pi\varepsilon_0} \dfrac{q}{r^2}$$

$$= 9 \times 10^9 \times \dfrac{10^{-8}}{\left(0.1\right)^2} = 9 \times 10^3 \ \text{N/C} \cdot$$

(ii)   At 8 cm form the centre i.e. point lies inside the sphere.

and E inside sphere = 0

(iii)   and field intensity at 1 m away from the centre

$$E = 9 \times 10^9 \times \frac{10^{-8}}{(1)^2} = 90 \text{ N/C}$$

Q.7 : Two spheres of radii 20 cm and 30 cm are charged separately with 40 esu and 60 esu of charge respectively. Find when they are connected (i) charge will flow form which sphere to which sphere (ii) Common potential.

Sol.: Here V_1 = Potential of sphere of radius 20 cm

$$= \frac{40}{20} = 2 \text{ esu}$$

V_2 = Potential of sphere of radius 30 cm

$$= \frac{60}{30} = 2 \text{ esu}$$

(i)   So charge will not flow.

(ii)   When they are connected, at equilibrium the

total charge = 40 + 60 = 100 esu

if the common potential be V then

$$V = \frac{q_1}{c_1} = \frac{q_2}{c_2}$$

so

$$\frac{q_1}{20} = \frac{100 - q_1}{30}$$

$$30 \, q_1 = 20 \times 100 - 20 \, q_1$$

$$30 \, q_1 = 20 \times 100$$

$$q_1 = 40$$

so

$$V = \frac{40}{20} = 2 \text{ esu}$$

Q.8: Evaluate if a metal sphere of radius 1 cm hold a charge of 1C? [WBUT - 2003]

Sol.: As

$$V = \frac{q}{C}$$

But for a spherical capacitor the capacitance

$$C = 4\pi\,\varepsilon_0 r$$

so
$$V = \frac{q}{4\pi\,\varepsilon_0 r} = 9\times10^9 \times \frac{1}{0.01}$$

$$= 9 \times 10^{11} \text{ Volts}$$

As the breakdown voltage for air is just 3kv/mm hence the surrounding air will be ionized and charge will leak to air.

Q.9 : The spherical region $0 < \gamma < 10$ contains a uniform volume charge density $\rho_V = 4\mu c/m^3$.

(i) Find Q_{enclosed} $0 < \gamma < 10$ cm.

(ii) Find D_r . $0 < \gamma < 10$ cm.   **[RU 2005]**

Sol.: Given

$$\rho_V = 4\mu c/m^3 \qquad\qquad 0 < \gamma < 10 \text{ cm.}$$

from Gauss law

$$\varepsilon_0 \oint \vec{E}.\,\overrightarrow{ds} = Q_{\text{encl.}} = \int_V \rho_V dV$$

(i)
$$Q_{\text{encl.}} = \int_V \rho_V dV = \rho_V \int dV$$

$$= \rho_V \int_{\phi=0}^{2\pi} \int_{\theta=0}^{\pi} \int_{r=0}^{10} r^2 \sin\theta \, dr \, d\theta \, d\phi$$

$$= \rho_V \frac{4}{3}\pi r^3 = 4\times10^{-6} \times \frac{4}{3}\pi r^3, \qquad 0 < r < 10 \text{ cm}$$

$$= 5.33\,\pi r^3\,\mu C \qquad 0 < r < 10 \text{ cm}$$

So
$$Q_{\text{encl.}} = 16.7456\,r^3\,\mu C$$

(ii)   as
$$\varepsilon_0 \oint_s \vec{E}.\,\overrightarrow{ds} = Q_{\text{encl.}} = \int_V \rho_V dV = \rho_V \int_V dV$$

$$\Rightarrow \qquad D_r\, 4\pi r^2 \;=\; \frac{4\pi r^3}{3}\,\rho_V$$

$$\Rightarrow \qquad \vec{D} \;=\; \frac{r}{3}\,\rho_V\,\hat{r}, \qquad 0 < r < 10 \text{ cm}$$

$$= \frac{r}{3}\,\rho_V\,\hat{r}$$

$$= 1.333\, r\,\mu C\hat{r}, \qquad 0 < r < 10 \text{ cm}.$$

Q.10 : Find the electric flux density $\vec{D}$ at a point A(6, 4, –5) caused by a confirm charge density $\rho_s = 60\ \mu C/m^2$ at a plane x = 8. [RU 2003, 2000]

Sol.: Given that

$$\rho_s \;=\; 60\ \mu C/m^2$$

as
$$\vec{A} \;=\; 6\hat{i} + 4\hat{j} - 5\hat{k}$$

$$\vec{D} \;=\; \varepsilon_0\,\vec{E} = \varepsilon_0\left(\frac{\rho_s}{2\varepsilon_0}\right)\hat{n}$$

$$= \frac{\rho_s}{2}\,\hat{n}$$

as plane is x = 8 but point is (6, 4, –5) where x coordinate is less than x = 8 so $\hat{n} = -\hat{i}$

$$\vec{D} \;=\; \frac{60\times 10^{-6}}{2}\left(-\hat{i}\right)$$

$$= -30\hat{i}\ \mu C/m^2.$$

Q.11 : In cylindrical coordinates (ρ, ϕ, z), electric flux density is given by $\vec{D} = ze\,\cos^2\phi\,\hat{z}$ c/m². Calculate the charge density at $\left(1, \dfrac{\pi}{4}, 3\right)$ and the total charge enclosed by the cylinder of radius 1 meter with $-2 \le z \le 2$ meter.

[RU 2001, 1999]

Sol.: as charge density is given by

$$\rho_V = \vec{\nabla}.\vec{D} = \frac{\partial\left(z\,\rho\,\cos^2\phi\right)}{\partial z}\hat{z}.\hat{z}$$

$$= \rho\,\cos^2\phi$$

$$\rho_V \text{ at } \left(1, \frac{\pi}{4}, 3\right) = \rho\,\cos^2\phi$$

$$= 1\cos^2\frac{\pi}{4}$$

$$= \frac{1}{2} = 0.5$$

and total charge

$$Q = \int_V \rho_V dV = \int_V \rho\,\cos^2\phi\,\rho\,d\rho\,d\phi\,dz$$

$$= \int_{z=-2}^{2} dz \int_{\phi=0}^{2\pi} \cos^2\phi\,d\phi \int_{\rho=0}^{1} \rho^2\,d\rho$$

$$= \frac{4}{3}\pi \text{ Coulombs}$$

Q.12 : For positive x, y and z, let $\rho_V = 40\ xyz\ c/m^2$. Calculate the total charge for the region defined by

(i) $0 \le x,\ y,\ z \le 2$

(ii) $0,\ y = 0,\ 0 \le 2x + 3y \le 10$ and $0 \le z \le 2$ [RU 2002]

Sol.: As

$$Q = \int_V \rho_V\,dV$$

(i) $$Q = \int_0^2 \int_0^2 \int_0^2 40\ xyz\ dx\ dy\ dz$$

$$= 40 \left[\frac{x^2}{2}\right]_0^2 \left[\frac{y^2}{2}\right]_0^2 \left[\frac{z^2}{2}\right]_0^2$$

$$= 40 \times 2 \times 2 \times 2 = 320 \text{ coulombs.}$$

(ii) Again

$$Q = \int_0^5 \int_0^{\frac{10-2x}{3}} \int_0^2 40 \text{ xyz dx dy dz}$$

$$= \int_0^5 40x \text{ dx} \left[\frac{y^2}{2}\right]_0^{\frac{10-2x}{3}} \left[\frac{z^2}{2}\right]_0^2$$

$$= \int_0^5 10x \text{ dx} \left(\frac{10-2x}{3}\right)^2$$

$$= \frac{40}{9} \int_0^5 \left(100x - 40x^2 + 4x^3\right) \text{dx}$$

$$= \frac{40}{9} \left[100\frac{x^2}{2} - 40\frac{x^3}{3} + 4.\frac{x^4}{4}\right]_0^5$$

$$= \frac{40}{9} \left[50 \times 25 - \frac{40}{3} \times 125 + 625\right]$$

$$= 925.93 \text{ Coulombs.}$$

Q.13 : Given the potential $V = \dfrac{10}{r^2} \sin \theta \cos \phi$. **Find electric flux density D at** $\left(2, \dfrac{\pi}{2}, 0\right)$.

[RU 2006]

Sol.: We know that

$$\vec{D} = \varepsilon_0 \, \vec{E}$$

$$\vec{E} = -\nabla V$$

$$= -\left[\frac{\partial V}{\partial r} \hat{r} + \frac{1}{r} \frac{\partial V}{\partial \theta} \hat{\theta} + \frac{1}{r \sin \theta} \frac{\partial V}{\partial \phi} \hat{\phi} \right]$$

$$= \frac{20}{r^3} \sin \theta \cos \phi \, \hat{r} - \frac{10}{r^3} \cos \theta \cos \phi \hat{\theta} + \frac{10}{r^3} \sin \phi \, \hat{\phi}$$

at $\left(2, \dfrac{\pi}{2}, 0 \right)$

$$\vec{D} = \varepsilon_0 \, \vec{E}$$

$$= \varepsilon_0 \left[\frac{20}{r^3} \hat{r} - \frac{10}{r^3}. 0 \, \hat{\theta} + \frac{10}{r^3}. 0 \, , \hat{\phi} \right]$$

as $\qquad\qquad r = 2$

$$D = 2.5 \, \varepsilon_0 \, \hat{r}$$

$$= 2212 \times 10^{-12} \hat{r} \ \ c/m$$

Q.14 : **Given the potential** $V = \dfrac{10}{r^2} \sin \theta \cos \phi$. **Find the work done in moving a** $-10 \, \mu c$ **charge from point A (1, 30°, 120°) to B(4, 90°, 60°).**

Sol.: The work done

$$W = -Q \int_{A}^{B} E \cdot dl = QV_{AB}$$

or $\qquad\qquad W = Q \left(V_B - V_A \right)$

$$= 10 \left[\frac{10}{16} \sin 90° \cos 60° - \frac{10}{1} \sin 30° \cos 120° \right] \times 10^{-6}$$

$$= 10\left[\frac{10}{32} - \frac{-5}{2}\right] \times 10^{-6}$$

$$= 28.125 \times 10^{-6} \text{ Joule}$$

Q.15 : Calculate electric field $\vec{E}$ at (20, 0) and (4, 6) for a potential value of V = $10y^3 + 20x^2$.

Sol.: As electric is given by

$$\vec{E} = -\vec{\nabla}V$$

$$= -\left(\frac{\partial}{\partial x}\hat{i} + \frac{\partial}{\partial y}\hat{j}\right)\left(10y^3 + 20x^2\right)$$

$$= -\left[\frac{\partial}{\partial x}\hat{i}\left(10y^3 + 20x^2\right) + \frac{\partial}{\partial y}\hat{j}\left(10y^3 + 20x^2\right)\right]$$

$$= -40x\,\hat{i} - 30y^2\hat{j}$$

Now at (20, 0)

$$E_{20,\,0} = -800\,\hat{i}$$

and at (4, 6)

$$E_{4,\,6} = -160\hat{i} - 108\hat{j}$$

Q.16 : Calculate the value of potential at any point inside and outside a uniformly charged sphere of radius 'R'.

Sol.:

Consider a sphere of radius 'R' and any point of consideration P at any distance r. Assume ρ_V be the volume charge density.

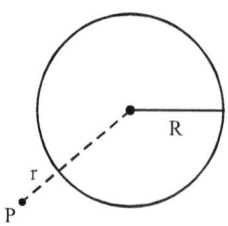

Fig.

When r > R

$$\vec{E} = \frac{Q}{4\pi\varepsilon_0 r^2}\hat{r} = \frac{e_V \int\limits_V dV}{4\pi\varepsilon_0 r^2}\hat{r}$$

$$= \frac{\rho_V\left(\frac{4}{3}\pi R^3\right)}{4\pi\varepsilon_0 r^2}\hat{r}$$

$$= \frac{\rho_V R^3}{3\varepsilon_0 r^2}\hat{r}$$

and electric potential

$$V = -\int\limits_0^r \vec{E}\cdot\vec{dl}$$

$$= -\int\limits_0^r \frac{\rho_V R^3}{3\varepsilon_0 r^2}(\hat{r}\cdot\hat{r})dr \qquad\qquad [\text{as } dl = \hat{r}\, dr]$$

$$= \frac{\rho_V R^3}{3\varepsilon_0 r}$$

when r < R

$$E = \frac{Q}{4\pi\varepsilon_0 r^2}\hat{r} \quad (0 < r < R)$$

$$= \frac{\rho_V\left(\frac{4}{3}\pi r^3\right)}{4\pi\varepsilon_0 r^2}$$

$$= \frac{\rho_V r}{3\varepsilon_0}$$

Electric potential $\qquad$ $V = -\int \vec{E} \cdot \vec{dl}$

$$= \left[\int_0^R \frac{\rho_V R^3}{3\varepsilon_0 r}(\hat{r}.\hat{r})\,dr - \int_R^r \frac{\rho_V r(\hat{r}.\hat{r})}{3\,\varepsilon_0}(\hat{r}.\hat{r})\,dr \right]$$

$$= \frac{\rho_V R^2}{3\,\varepsilon_0} - \frac{\rho_V r^2}{6\varepsilon_0} + \frac{\rho_V R^2}{3\varepsilon_0}$$

$$= \frac{\rho_V R^2}{3\,\varepsilon_0}\left[3R^2 - r^2 \right]$$

Q.17 : When electric field is always directed towards X-direction then prove that potential will be independent of Y and Z coordinates. Also prove that when field is constant, there will not be any free charge in that region.

[WBUT - 2007]

Sol.: According to the questions

$\qquad$ $\vec{E} = \vec{E}_x \hat{i}$ $\qquad\qquad$ ...(1)

and $\qquad$ $\vec{E} = -\vec{\nabla} V$

$$= -\frac{\partial V_x}{\partial x}\hat{i} - \frac{\partial V_y}{\partial y}\hat{j} - \frac{\partial V_z}{\partial z}\hat{k} \qquad\qquad ...(2)$$

Comparing (1) and (2)

$$\frac{\partial V_y}{\partial y} = 0 \quad \text{and} \quad \frac{\partial V_z}{\partial z} = 0$$

so $\qquad$ $V_y = V_z = \text{constant}$

Again $\qquad$ $\vec{E} = -\vec{\nabla} V$ shows that when

$\qquad$ $\vec{E} = \text{constant} \rightarrow \vec{\nabla} V = \text{constant}$

So $\qquad$ $V(x, y, z) = 0$

and form poisson's relation

$$\nabla^2 V = \frac{\rho}{\varepsilon_0}$$

or $$\rho = \varepsilon_0 \ \nabla^2 V = 0$$

i.e. it is charge free region.

Q.18 : A long cylinder carries charge proportional to the distance form the axis (r). If the cylinder is with radius R, then find the electric field both at $r > R$ and $r < R$ using Gauss's law of electrostatics.

Sol.: Consider a long cylinder of radius R and height h.

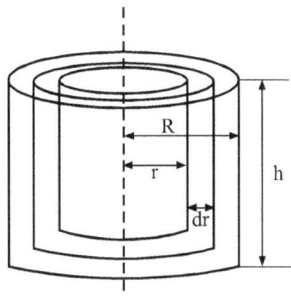

Fig.

so the charge enclosed

$$Q = \int_0^R (kr)(2\pi \, rdr)h$$

$$= \frac{k \, 2\pi h. \, R^3}{3}$$

Case - I : if $r > R$ then

$$\int \vec{E}. \, \vec{ds} = \frac{Q}{\varepsilon_0}$$

or $$E \, 2\pi \, rh = \frac{k. \, 2\pi h. \, k^3}{3\varepsilon_0}$$

$$E = \frac{k\,R^3}{3\varepsilon_0 r}$$

Case - II : When $r < R$ then

$$q' = \int_r^h (kr)\,(2\pi\,rh)\,dr$$

$$= \frac{k.2\pi h\;r^3}{3}$$

so $\qquad\qquad E = \dfrac{kr^3}{3\varepsilon_0 r} = \dfrac{kr^2}{3\varepsilon_0}$

Q.19 : Verify that potential function $V_{x,y} = 3x^2 + 2y^2$ satisfies Laplace's equation or not. Also find the charge density.

Sol.: Laplace's equations states that

$$\nabla^2 V = 0 \text{ for a charge free region}$$

here $\qquad\qquad V = 3x^2 + 2y^2$

$$\nabla^2 V = \frac{\partial^2}{\partial x^2}\left(3x^2 + 2y^2\right) + \frac{\partial^2}{\partial y^2}\left(3x^2 + 2y^2\right)$$

$$= 6 + 4 = 10$$

$$\neq 0$$

as $\nabla^2 V \neq 0$ it shows that potential function does not satisfy Laplace equation and because it is not a charge free region here poisson's equation will hold true

where $\qquad\qquad \nabla^2 V = -\dfrac{\rho}{\varepsilon_0}$

here $\qquad\qquad -\dfrac{\rho}{\varepsilon_0} = 10$

so $\rho = -10 \, \varepsilon_0 \; c/m^3$

Q.20 : A condenser consisting of an oxidized aluminimum sheets with effective surface area of 400 cm² with a capacitance of 8μF. Calculate

(i)   The field strength if ε_r = 8 and potential diff an between aluminium and the electrolyte = 10 V.

(ii)   Total dipole moment induced in the oxide layer and its susceptibility.

Sol.: Here

$$A = 400 \text{ cm}^2 = 400 \times 10^{-4} \text{ m}^2$$

$$\varepsilon_r = 8, \; C = 8 \; \mu F, \; V = 10 \text{ V}$$

Electric field $E = \dfrac{V}{d}$

and $d = \dfrac{\varepsilon_0 \, \varepsilon_r A}{C}$

$$= \frac{8.854 \times 10^{-12} \times 8 \times 400 \times 10^{-4}}{8 \times 10^{-6}}$$

$$= 3541 \times 10^{-10} \text{ m.}$$

(i) $E = \dfrac{V}{d} = \dfrac{10}{3541 \times 10^{-10}} = 2.8 \times 10^7 \text{ V/m}$

(ii)   Polarization $P = \varepsilon_0 E \left(\varepsilon_r - 1 \right)$

$$= 8.854 \times 10^{-12} \times 2.8 \times 10^7 \left(8 - 1 \right)$$

$$= 1.7354 \times 10^{-3}$$

and $P = \dfrac{p}{\text{volume}}$

so $p = p \times \text{volume} = p \times (A \times d)$

$$= 1.7354 \times 10^{-3} \times 400 \times 10^{-4} \times 3541 \times 10^{-10}$$

$$= 2.5 \times 10^{-11}$$

and susceptibility

$$\chi = \frac{p}{\varepsilon_0 E}$$

$$= \frac{2.5 \times 10^{-10}}{8.854 \times 10^{-12} \times 2.8 \times 10^7}$$

$$= 0.1 \times 10^{-6}$$

Q.21 : A dielectric material contains 2×10^9 polar molecules/m^3 each of dipole moment 1.8×10^{-27} cm. Assuming that all of the dipoles are aligned towards electric field $E = 10^5$ V/m. Find the polarization, electric susceptibility and the relative permittivity.
[IES 97]

Sol.: It is given that

No. of molecules $N = 2 \times 10^9$ moleucles/m^3

Dipole moment $p = 1.8 \times 10^{-27}$ c-m

Electric field $E = 10^5$ V/m.

Hence polarization

$$P = Np$$

$$^- 2 \times 10^9 \times 1.8 \times 10^{-27}$$

$$= 3.6 \times 10^{-18} \ c/m^2$$

susceptibility

$$\chi = \frac{P}{\varepsilon_0 E} = \frac{3.6 \times 10^{-18}}{8.854 \times 10^{-12} \times 10^5}$$

$$= 4.07 \times 10^{-12}$$

and relative permittivity

$$\varepsilon_r = 1 + \chi = 1 + 4.07 \times 10^{-12}$$

$$\sim 1$$

Q.22 : A certain homogenous slab of loss-less dielectric material is characterized by an electric susceptibility of 0.12 and carries a uniform flux density within it of 1.6 n c/m^2. Calculate the electric field intensity, the polarization, and the average dipolemoment. Given No. of dipoles per cubic meter $= 2 \times 10^{19}$ and separation between two equipotential surfaces $= 2.54$ cm.
[IES - 1998]

Sol.: Electrical susceptibility

$$\chi_e = 0.12, \ D = 1.6 \ n \ C/m^2$$

We know

$$\varepsilon_r = 1 + \chi_e$$
$$= 1 + 0.12$$
$$= 1.12$$

$$D = \varepsilon_0 \, \varepsilon_r E$$

$$E = \frac{D}{\varepsilon_0 \, \varepsilon_r} = \frac{1.6 \times 10^{-9}}{8.854 \times 10^{-12} \times 1.12}$$

$$= 161.34 \ V/m$$

and voltage

$$V = E \times \text{separation}$$
$$= 161.34 \times 2.54 \times 10^{-2}$$
$$= 4$$

$$\text{Polarization } P = \chi_e \, \varepsilon_0 E$$
$$= 0.12 \times 8.854 \times 10^{-12} \times 161.34$$
$$= 0.1714 \ n \ c/m^2$$

and   Average dipole moment $= \dfrac{0.1714 \times 10^{-9}}{2 \times 10^{19}}$

$$= 8.8714 \times 10^{-30} \ \text{c-m.}$$

Summary

- Vector form of Coulomb's law

$$F = \frac{1}{4\pi\varepsilon_0} \frac{q_1 q_2}{|\vec{r}|^2} \hat{r}$$

- Linear charge density

$$\lambda = \lim_{\Delta l \to 0} \frac{\Delta q}{\Delta l} \ c/m$$

surface charge density

$$\sigma = \lim_{\Delta l \to 0} \frac{\Delta q}{\Delta l} \ c/m^2$$

Volume charge density

$$l = \lim_{\Delta l \to 0} \frac{\Delta q}{\Delta l} \ c/m^3$$

- Electric field strength

$$\vec{E} = \frac{1}{4\pi\varepsilon_0} \frac{q}{r^2} \hat{r} \ N/c$$

- Electric displacement vector

$$\vec{D} = \varepsilon_m \vec{E}$$

- Electrostatic potential energy

$$U = \frac{1}{4\pi\varepsilon_0} \frac{q_1 q_2}{r_{12}}$$

- Electrostatic potential is electrostatic potential energy per unit positive test charge

$$V = \frac{U}{q_2} = \frac{1}{4\pi\varepsilon_0} \frac{q_1}{r_{12}}$$

- Electric flux

$$\phi = \oint_s \vec{E} \cdot \vec{ds}$$

- Gauss law of electrostatic

$$\oint_s \vec{E} \cdot \vec{ds} = \frac{Q_{enclosed}}{\varepsilon_0}$$

or $$\vec{D} \oint_s \vec{ds} = Q_{enclosed}$$

- Differential form of Gauss's law

$$\vec{\nabla} \cdot \vec{E} = \frac{\rho}{\varepsilon_0}$$

or $\vec{\nabla} \cdot \vec{E} = \rho$

- Electric field $\vec{E} = -\vec{\nabla} V$
- Poisson's equation

$$\nabla^2 V = 0$$

- Equation of contivity

$$\vec{\nabla} \cdot \vec{J} + \frac{\partial \rho}{\partial t} = 0$$

- Drift velocity

$$v_d = \frac{eE}{m} \tau \text{ where } \tau \text{ is relaxation time.}$$

- Electric current

$$I = ne A vd$$

- Current density

$$\vec{J} = \sigma \vec{E}$$

- Dipole moment

$$\vec{p} - q\vec{r}$$

- Polarization $\vec{p} = \chi \varepsilon_0 \vec{E}$

-

$$\vec{\Delta} = \varepsilon_0 \vec{E} + \vec{P}$$

- Dielectric constant or relative permittivity

$$\varepsilon_r = k = 1 + \chi$$

EXERCISE

1.   Give statement of coulomb's law of electrostatics calculate the electric field intensity and electric potential for any charge.

2. What do you understand by electric flux and velocity flux?

3. Give statement of Gauss' law of electrostatic. Give its differential form also.

4. Using Gauss's law of electrostates, derive Coulomb's law.

5. What is electrostatic potential energy and electrostatic potential?

6. From the differential form of Gauss' law, find poisson's and Laplace's equations.

7. For any electric field $\vec{E}$ prove that $\vec{\nabla} \times E = 0$

8. For any electric field prove that $\vec{E} = -\vec{\nabla}V$ where V is potential.

9. Find electro field intensity at any point inside and outside for a cylindrical charge distribution.

10. Explain the behaviour of a dielectric material placed in an electric field and hence explain electric polarization of matter.

11. What are polar and non-polar molecules? Discuss different types of polarizations in dielectrics.

12. What do you understand by atomic polarizability? Fin a relation between dipole moment and atomic polarizability.

13. Explain the terms permittivity, dielectric coefficients, susceptibility and dielectric polarization. Derive their relation also.

14. Show that $\vec{D} = \varepsilon_0 \vec{E} + \vec{P}$

15. Define the terms dielectric constant and electrical susceptibility and prove that
$$k = 1 + x_e.$$

16. State and prove Gauss's law in dielectrics.

17. For any electric potential given by $V(x,y,z) = \left(2x^2 + 4y^2 + 3z^2\right)$ find the electric field at (1, 1, 1). [WBUT 2005]

18. In space the electric field is given by $\vec{E} = 8\hat{i} + 4\hat{j} + 3\hat{k}$. Calculate the electric flux through a surface area 100 sq. units in xy planes. [WBUT 2004]

3

MAGNETOSTATICS AND TIME VARYING FIELDS

Contents: Lorentz forces, force on a small current element placed in a magnetic field, Biot-Savart-law and its applications, divergence of magnetic field, vector potential, Ampere's law in integral form and conversion to differential form. Faraday's law of electromagnetic induction in integral form and conversion to differential form.

FORCE ON A CURRENT CARRYING CONDUCTOR IN A MAGNETIC FIELD

When a current carrying conductor is placed in a magnetic field, a magnetic force works on it in perpendicular direction to both current and magnetic field.

Consider a portion of length l and cross sectional area A of a conductor carries a current I placed in any magnetic field $\vec{B}$ directed downward into the page.

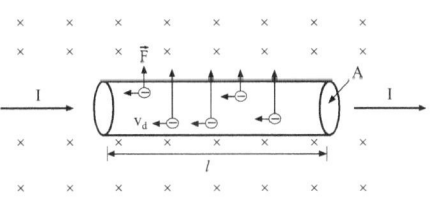

As current in the conductor is due to free electron drift from lower to higher potential ends of the conductor, magnetic force works on these electrons. If there are n electrons per unit volume then from

$$F = q\,\vec{v} \times \vec{B}$$

for each electrons $\qquad F' = e\,v_d B \qquad\qquad\qquad\qquad ...(i)$

No. of electrons for 1 length

$$N = n \, Al$$

so total force

$$F = F'N$$

$$= \left(e v_d B\right)\left(n \, Al\right)$$

$$= \left(n e A v_d\right) Bl$$

$$F = I \, Bl \qquad \left(\text{as } I = neAvb_d\right)$$

if θ is the angle between conductor and $\vec{B}$ then

$$\boxed{F = IBl \sin \theta} \qquad \qquad ...(ii)$$

or

$$\boxed{\vec{F} = I \, \vec{l} \times \vec{B}} \qquad \qquad ...(iii)$$

where $\vec{l}$ is a displacement vector in the direction of current.

LORENTZ FORM

It has been observed that stationary electric charge does not have any effect on the nearby magnet shown no evidence of any magnetic field when charge is at rest, but H.C. Oersted found that when any wire with current in it is placed near to any magnetic compass, the needle of compass deflects as soon as their is the movement of charge, showing presence of magnetic field when there is current i.e. charge is in motion. A compass needle experiences a force, when placed near to a current carrying conductor.

Since a current has moving electric charges, that means moving electric charge is responsible for generation of magnetic field around the conductor causing compass needle to deflect. It clearly shows that free electrons or freely moving charged practices (not in the wire) would also experience a force when passing through a magnetic field.

Let us assume that there are N charged particles of charge q pass by a given point in time t, then there is a current

$$I = \frac{Nq}{t}$$

Suppose t be the time for a charge to travel a distance l in presence of a magnetic field $\vec{B}$. Then

$$\vec{l} = \vec{v} \, t, \text{ where } \vec{v} \text{ is the velocity of charged particle.}$$

The force on these N particles

$$\vec{F} = I\vec{l} \times \vec{B} = \left(\frac{Nq}{t}\right)(\vec{v}t) \times \vec{B}$$

$$= Nq\,\vec{v} \times \vec{B}$$

so force on any one particle is

$$\vec{F} = q\vec{v} \times \vec{B} \qquad\qquad ...(i)$$

when an electric charge is at rest or move, in any electrostatic field $\vec{E}$, then the electrostatic force $= q\vec{E}$ ....(ii)

This combined effect of force experienced by a charged particle moving in an electric and magnetic field is known as lorentz forces.

so if charge q moves with a velocity $\vec{v}$ in a combined electric and magnetic field, then the total force, known as Lorentz force is given by

$$\vec{F} = q\,\vec{E} + q\left(\vec{v} \times \vec{B}\right)$$

$$\boxed{\vec{F} = q\left[\vec{E} + \vec{v} \times \vec{B}\right]} \qquad\qquad ...(iii)$$

The above equation is also known as Lorentz equation for total electromagnetic force on any charge moving in an electric field along with a magnetic field.

Here magnetic Lorentz force varies directly with the strength of magnetic field $\vec{B}$ as well as with velocity $\vec{v}$. So if charged particle stops i.e. v = 0 so magnetic lorentz force becomes zero. the magnetic Lorentz force does not work on charge because its direction is normal to the charge motion. As $F = q\left(\vec{v} \times \vec{B}\right)$ which shows that Lorentz magnetic force is in the direction of $\vec{v} \times \vec{B}$ i.e. perpendicular to both direction of charge motion $\left(\vec{v}\right)$ and strength of magnetic field.

SOME IMPORTANT TERMS

Magnetic flux

Magnetic flux is defined as the total number of magnetic field lines passing through a surface in the normal direction.

Mathematically it is equal to

$$\phi_B = \oint_s \vec{B}.\,\overline{ds} \qquad\qquad\text{Its unit is weber (wb)}$$

Magnetic flux density $(\vec{B})$ or magnetic induction $(\vec{B})$

It is defined as the magnetic flux per unit area. Its SI unit in wb/m² or Tesla (T).

Magnetic Intensity $(\vec{H})$

It is defined as the ratio of magnetic flux density $(\vec{B})$ to the permeability(μ) of the medium.

so
$$\vec{H} = \frac{\vec{B}}{\mu}$$

Its SI unit is A/m.

Biot-Savart Law

Experimentally it has been verified that a current carrying conductor can produce a magnetic field around it. The direction and magnitude of this magnetic field can be determined using Biot-Savrt law, according to which "the magnetic induction $d\vec{B}$ at any point P due to an element of infinitesimal length $\vec{dl}$ with current I is proportional to (a) the current element dl. (b) the current I through dl (c) sine of angle θ between the element dl and radius vector r. and (d) inversely proportional to the radius vector r which is the distance of point P from dl.

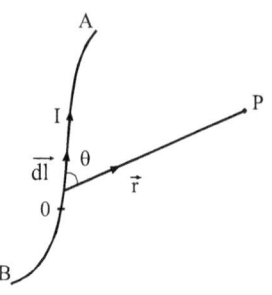

i.e.
$$dB \propto \frac{I\,dl\,\sin\theta}{r^2}$$

$$d\vec{B} = \frac{\mu_0}{4\pi}\frac{Idl\,\sin\theta}{r^2} \qquad \text{...(i)}$$

where μ_0 is the permeability of free space.

and
$$\mu_0 = 4\pi \times 10^{-7}\ \text{wb A}^{-1}\text{m}^{-1}$$

$$\vec{dB} = \frac{\mu_0}{4\pi}I\frac{\vec{dl}\times\vec{r}}{r^3} = \frac{\mu_0}{4\pi}I\frac{\vec{dl}\times\hat{r}}{r^2} \qquad \text{...(ii)}$$

equation (ii) is the vector form of Bio-Savart law.

Now if

$$\theta = 0° \rightarrow \left|d\vec{B}\right| = 0$$

which shows that magnitude of magnetic flux density becomes zero at any axial point on the current carrying conductor.

if $\theta = 90°$ then

$$\left|d\vec{B}\right| = \frac{\mu_0}{4\pi} \frac{I dl}{r^2}$$

it shows the maximum value of magnetic field intensity at any point on perpendicular line to the current carrying length element.

The total magnetic field intensity due to complete length of current carrying conductor

$$\vec{B} = \int d\vec{B}$$

$$\vec{B}(\vec{r}) = \frac{\mu_0 I}{4\pi} \int \frac{\overrightarrow{dl} \times \hat{r}}{r^2}$$

$$\boxed{\vec{B}(\vec{r}) = \frac{\mu_0 I}{4\pi} \int \frac{\overrightarrow{dl} \times \vec{r}}{r^3}} \qquad ...(iii)$$

Equation (iii) is the integral form of biot-Savart's law which is analogous to Coulomb's law of electrostatics. The current element I dl can be represented in terms of surface current density J_s or volume current density J_V as

$$I \overrightarrow{dl} = J_s \, ds$$

and $\qquad I \, dl = J_V \, dV$

and hence Biot-Savart's law can accordingly be written as

$$\vec{B} = \frac{\mu_0}{4\pi} \int_s \frac{\vec{J}_s \times \vec{r}}{r^3} \overrightarrow{dS} \quad \text{(surface current distribution)} \qquad ...(iv)$$

and for volume current distribution

$$\vec{B} = \frac{\mu_0}{4\pi} \int_V \frac{\vec{J}_V \times \vec{r}}{r^3} \overrightarrow{dV} \qquad ...(v)$$

APPLICATIONS OF BIOT-SAVRT'S LAW

(i)   Magnetic field at the centre of a current carrying circular coil

Suppose there is a circular coil of radius r, carrying a current I as shown

According to Biot-Savart law, the magnitude of magnetic field at O due to a small element dl of the coil is

here

$$dB = \frac{\mu_0}{4\pi} \frac{I\,dl\,\sin\theta}{r^2}$$

But $\theta = 90°$

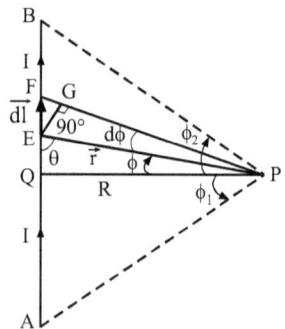

so $$dB = \frac{\mu_0}{4\pi} \frac{I\,dl}{r^2}$$

and total field due to entire coil

$$B = \frac{\mu_0}{4\pi} \frac{I}{r^2} \int dl$$

$$= \frac{\mu_0}{4\pi} \frac{I}{r^2} (2\pi r) \left(\text{as } \int dl = 2\pi r \right)$$

$$\boxed{B = \frac{\mu_0}{2} \frac{I}{r}}$$

Magnetic field B is perpendicular to the plane of the coil, directed upwards. If the current is reversed the field will be downward directed.

(ii)   Magnetic field due to a straight current carrying conductor of finite length

Suppose AB is a straight conductor carrying a current of I and magnetic field intensity is to be determined at point P.

According to Biot-Savart law the magnetic field at P

$$\overrightarrow{dB} = \frac{\mu_0}{4\pi} \frac{I\,\overrightarrow{dl} \times \overrightarrow{r}}{r^3}$$

angle between $I\,\overrightarrow{dl}$ and $\overrightarrow{r}$ is $(180 - \theta)$ so

$$dB = \frac{\mu_0}{4\pi} \frac{I\,dl\,\sin(180-\theta)}{r^2}$$

$$dB = \frac{\mu_0}{4\pi} \frac{I\,dl\,\sin\theta}{r^2} \qquad ...(i)$$

Now $\qquad$ EG = EF sin θ

$$= dl \sin \theta$$

and $\qquad$ EG = EP sin dφ = r sin dφ

$$= r \, d\phi$$

so $\qquad$ dl sin θ = r dφ $\qquad$ ...(ii)

so from (i)

$$dB = \frac{\mu_0}{4\pi} \frac{I \, d\phi}{r} \qquad \text{...(iii)}$$

from Δ EQP, $\qquad$ $r = \dfrac{R}{\cos \phi}$

so $\qquad$ $dB = \dfrac{\mu_0}{4\pi} \dfrac{I \cos\phi \, d\phi}{R}$ $\qquad$ ...(iv)

Then the total magnetic field at point P due to the entire conductor is

$$B = \int_{-\phi_1}^{\phi_2} \frac{\mu_0}{4\pi} \frac{I}{R} \cos \phi \, d\phi$$

$$= \frac{\mu_0}{4\pi} \frac{I}{R} \left[\sin \phi \right]_{-\phi_1}^{\phi_2}$$

$$\boxed{B = \frac{\mu_0}{4\pi} \frac{I}{R} \left(\sin \phi_1 + \sin \phi_2 \right)} \qquad \text{...(v)}$$

for any conductor of infinite length

$$\phi_1 = \phi_2 = 90°$$

so $\qquad$ $B = \dfrac{\mu_0}{4\pi} \dfrac{2I}{R}$

$$\boxed{B = \frac{\mu_0 I}{2\pi R}} \, NA^{-1} \, m^{-1}$$

The direction of magnetic field due to a current carrying conductor can be obtained by using any of the laws like

(i)   Right hand palm rule no 1
(ii)   Right hand thumb rule or
(iii)   Maxwell Right-hand screw Rule.

(iii)   Force between two parallel current carrying conductors and definition of ampere

Experimentally it has been observed that two current carrying conductors attract each other when the currents in them are in same direction and repel each other when these are in opposite directions.

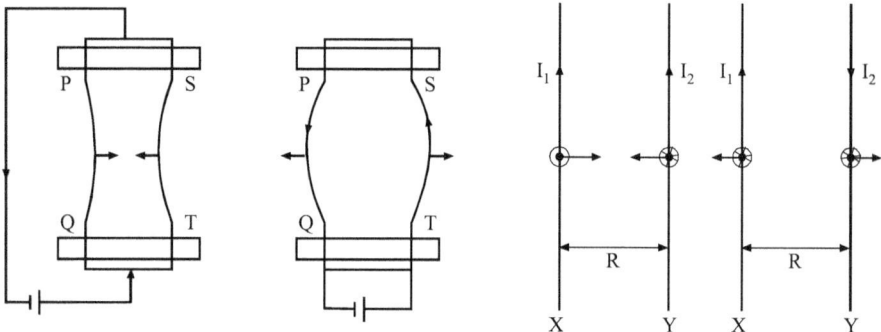

Suppose X and Y are two long parallel straight conductors with currents I_1 and I_2 amp. respectively. The magnitude of the magnetic field $\vec{B}$, at any point on Y due to the current I_1 in X is given by

$$B_1 = \frac{\mu_0 \, I_1}{2\pi \, R}$$

Perpendicular to plane of the page directed downward.

Hence current carrying conductor Y is thus situated in a magnetic field $\vec{B}_1$ perpendicular to its length. Hence it experiences a magnetic force, magnitude of which is given by

$$F = I_2 \, B_1 l$$

$$= I_2 \left(\frac{\mu_0 I_1}{2\pi R} \right) l$$

and force per unit length

$$\boxed{\frac{F}{l} = \frac{\mu_0}{2\pi} \, \frac{I_1 I_2}{R}}$$

according to Fleming left hand rule the direction of this force is towards X when I_1 & I_2 are in same direction and is away from X when I_1 & I_2 are in opposite direction.

Similarly force per unit length of X due to current in Y is $\dfrac{\mu_0}{2\pi}\dfrac{I_1 I_2}{R}$, directed opposite to forces on Y due to X.

The directions of the forces on the two conductors show that the conductors attract each other if currents are in same direction and repel each other if currents are in opposite direction.

If
$$I_1 = I_2 = I$$

so
$$\frac{F}{1} = \frac{\mu_0}{2\pi}\frac{I^2}{R}$$

If now the value of current is such that when R = 1 m, $\dfrac{F}{1} = 2\times10^{-7}$ N/m , then the current

is said to be one ampere. So one ampere is that current which can produce a force of 2×10^{-7} N/m in vacuum between two infinitely long conductors placed 1 meter apart.

So
$$2\times10^{-7}\,\frac{N}{m} = \frac{\mu_0}{2\pi}\frac{(1A)^2}{1m}$$

so
$$\frac{\mu_0}{4\pi} = 10^{-7}\,\frac{N}{A^2}\left[=10^{-7}\,\frac{Wb}{A-m}\right]$$

(iv)   Magnetic field at the axis of a current carrying circular coil

Suppose there is a circular coil of radius a carrying a current I

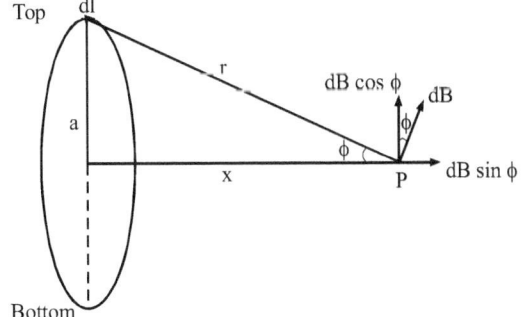

According to Biot-Savart law, the magnitude of magnetic field due to the small current element at P will be

$$dB = \frac{\mu_0}{4\pi} \frac{I \, dl \sin \theta}{r^2}$$

here θ is the angle between length element and the line joining to the point P. i.e. 90°.

so
$$dB = \frac{\mu_0}{4\pi} \frac{Idl}{r^2} \qquad \qquad ....(i)$$

The direction will be right angle to the line joining length element to point P. This can be resolved into two components dB sin ϕ along the axis of coil and other dB cos ϕ at right angle to axis. The components which are along the axis will be added up. All the vertical components are equal and opposite and cancel each other.

Hence the resultant magnetic field B at P is directed along the axis is

$$B = \int dB \sin \phi$$

$$= \frac{\mu_0 I}{4\pi \, r^2} \int dl \sin \phi$$

here
$$\sin \phi = \frac{a}{r}$$

so
$$B = \frac{\mu_0}{4\pi} \frac{Ia}{r^3} \int dl$$

But
$$\int dl = 2\pi a$$

and
$$r^2 = a^2 + x^2$$

so
$$B = \frac{\mu_0}{4\pi} \frac{2\pi \, Ia^2}{\left(a^2 + x^2\right)^{3/2}}$$

$$= \frac{\mu_0 Ia^2}{2\left(a^2 + x^2\right)^{3/2}}$$

if the coil has N turns then

$$B = \frac{\mu_0}{4\pi} \frac{2\pi N Ia^2}{\left(a^2 + x^2\right)^{3/2}} = \frac{\mu_0 N Ia^2}{2\left(a^2 + x^2\right)^{3/2}} NA^{-1}m^{-1}$$

direction of this magnetic field will be along the axis.

klif x = 0 i.e. at the centre of coil

$$B = \frac{\mu_0}{4\pi} \frac{2\pi \, NI}{a}$$

$$B = \frac{\mu_0 NI}{2a} \, NA^{-1}m^{-1}$$

(v)   Magnetic field of a solenoid

A solenoid is considered to be a long cylindrical helix.

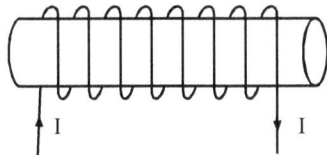 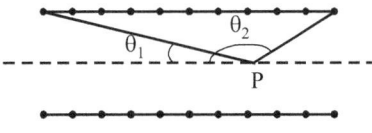

Figure shows a solenoid of radius a meter and carrying a current I ampere. The lines of forces inside the solenoid are nearly parallel which clearly represents that the magnetic field within the solenoid is uniform and parallel to the axis of solenoid.

Suppose there are n turns per unit length of the solenoid. Consider solenoid to be divided up into a number of small coils and consider one of those coils is AB of dx width.

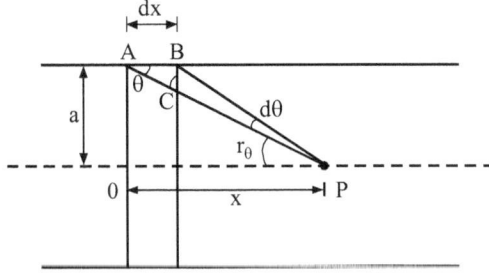

Fig.

The number of turns in this coil is n dx.

The magnetic field at P due to this small coil is

$$dB = \frac{\mu_0 (n\,dx) I \, a^2}{2(a^2 + x^2)^{3/2}} \, NA^{-1}m^{-1}$$

according to the figure

$$\sin \theta = \frac{BC}{AB} = \frac{r d\theta}{dx}$$

or

$$dx = \frac{r d\theta}{\sin \theta}$$

and

$$r^2 = a^2 + x^2$$

Hence

$$dB = \frac{\mu_0 (nr\, d\theta)\, I a^2}{2r^3 \sin \theta}$$

$$= \frac{\mu_0 n\, I a^2}{2r^2} \frac{d\theta}{\sin \theta}$$

and

$$\sin \theta = \frac{a}{r}$$

or

$$a^2 = r^2 \sin^2 \theta$$

so

$$dB = \frac{\mu_0 n\, I}{2} \sin \theta\, d\theta$$

The total magnetic field at P can be obtained between the first and last turn of the solenoid i.e. between the limit of θ_1 and θ_2.

so

$$B = \int_{\theta_1}^{\theta_2} dB = \int_{\theta_1}^{\theta_2} \frac{\mu_0 n\, I}{2} \sin \theta\, d\theta$$

$$= \frac{\mu_0 n\, I}{2} \left[-\cos \theta \right]_{\theta_1}^{\theta_2}$$

$$\boxed{B = \frac{\mu_0 n\, I}{2} \left(\cos \theta_1 - \cos \theta_2 \right)} \quad NA^{-1}\ m^{-1}$$

If P lies will inside the solenoid

i.e.

$$\theta_1 \approx 0 \quad \text{and} \quad \theta_2 \approx 180° \quad \text{so}$$

$$B = \frac{\mu_0 n\, I}{2} \left[1 - (-1) \right]$$

$$\boxed{B = \mu_0 n I} \quad NA^{-1}\ m^{-1}$$

at the ends of the solenoid

$$\theta_1 = 0$$

and $$\theta_2 = 90°$$

so $$\boxed{B = \frac{\mu_0 nI}{2}} NA^{-1}m^{-1}$$

which clearly shows that the magnetic field at the ends of a solenoid is half of that at the centre of the solenoid.

for a very long solenoid, field is almost uniform and is parallel to the solenoidal axis.

Gauss's law in magnetism (Divergence of magnetic field)

We know that divergence of any vector field can be defined as the limiting value of ratio of flux across any closed surface around the point to the volume when volume is contracted to zero.

i.e. $$\vec{\nabla}\cdot\vec{B} = \lim_{\Delta V \to 0} \frac{\oint_s \vec{B}.\overrightarrow{ds}}{\Delta V} \qquad ...(i)$$

The magnetic flux through a closed surface is always zero

i.e. $$\oint_s \vec{B}.\overrightarrow{ds} = 0 \qquad ...(ii)$$

which is Gauss's law in magnetism.

From Gauss's Divergence theorem

$$\oint_s \vec{B}.\overrightarrow{ds} = \int_V \left(\vec{\nabla}\cdot\vec{B}\right) dV \qquad ...(iii)$$

from (ii) & (iii)

$$\int_V \left(\vec{\nabla}\cdot\vec{B}\right) dV = 0 \qquad ...(iv)$$

Hence $$\vec{\nabla}\cdot\vec{B} = 0 \qquad ...(v)$$

Equation (v) shows that free isolated magnetic poles does not exist i.e. magnetic poles always exist in pairs.

Vector Potential

If the divergence of any vector is zero, then there exists a vector $\vec{A}$ whose curl gives the given vector. Such a vector is called vector potential of that vector field.

As $\text{div}\left(\text{curl } \vec{A}\right) = 0$ then $\vec{A}$ is the vector field.

In magnetostatic, the divergence of magnetic field induction $\vec{B}$ is zero,

i.e., $$\vec{\nabla} \cdot \vec{B} = 0 \qquad \qquad \text{...(i)}$$

and $$\vec{\nabla} \cdot \left(\vec{\nabla} \times \vec{A}\right) = 0 \qquad \qquad \text{...(ii)}$$

By comparing (i) and (ii)

$$\vec{B} = \vec{\nabla} \times \vec{A} \qquad \qquad \text{...(iii)}$$

hence here $\vec{A}$ can be the vector potential of magnetic field $\vec{B}$.

Hence any magnetic vector potential is defined as a vector function, the curl of which gives the magnetic field induction.

In magnetostatic vector potential can be written as

$$\vec{A} = \frac{\mu_0}{4\pi} \int_V \frac{\vec{J}\,dV}{r} \quad \text{Joules/Amp. or JA}^{-1}$$

here V is the volume of the source surface.

As biot-Savart law in terms of vector potential is very simple hence vector potential $\left(\vec{A}\right)$ can easily be calculated and then magnetic field induction $\left(\vec{B}\right)$ also can be found by taking curl of $\vec{A}$.

Ampere's circuital law

According to ampere's circuital law the line integral of magnetic field $\vec{B}$ around any closed curve is equal to μ_0 times the net current i passing through the area enclosed by the closed curve.

i.e. $$\oint \vec{B} \cdot \vec{dl} = \mu_0 i$$

where μ_0 is free space permeability.

Proof : Consider AB as along, straight conductor with current i, as shown

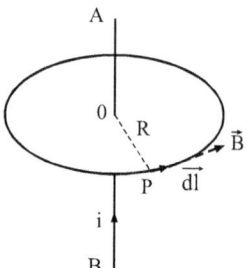

according to **Biot-Savart law, magnetic field at P,**

$$B = \frac{\mu_0}{2\pi} \frac{i}{R} \qquad \text{...(i)}$$

and at P line integral

$$\oint \vec{B} \cdot \vec{dl} = \oint B \, dl = B \oint dl \qquad \text{...(ii)}$$

Using (i) and $\oint dl = 2\pi R$ in (ii)

$$\oint \vec{B} \cdot \vec{dl} = \frac{\mu_0 i}{2\pi R} (2\pi R)$$

$$= \mu_0 i$$

which is Ampere's circuital law.

This is the integral form of Ampere's circuital law.

Conversion to Differential form

As enclosed current I can be stated as

$$I = \oint_s \vec{J} \cdot \vec{ds} \qquad \text{...(i)}$$

where $\vec{j}$ is the current density and $\vec{ds}$ is the small surfaces area of closed path.

From Ampere's circuital law

$$\oint \vec{B} \cdot \vec{dl} = \mu_0 I$$

$$= \mu_0 \oint_s \vec{J} \cdot \vec{ds} \qquad \text{...(ii)}$$

Stoke's law states that

$$\oint_s (\nabla \times B) \cdot \vec{ds} = \oint B \cdot \vec{dl} \qquad \text{...(iii)}$$

from (ii) and (iii)

$$\oint_s (\vec{\nabla} \times \vec{B}) \cdot \vec{ds} = \mu_0 \oint \vec{J} \cdot \vec{ds}$$

hence

$$\boxed{\vec{\nabla} \times \vec{B} = \mu_0 \vec{J}} \qquad \text{...(iv)}$$

equation (iv) is the differential form of Amperes law. Because $\vec{\nabla} \times \vec{B} \neq 0$, magnetic field is not conservative and its curl has some value.

When the points are inside a closed loop for which $\vec{J} = 0$, $\vec{\nabla} \times \vec{B} = 0$.

APPLICATIONS OF AMPERE'S CIRCUITAL LAW

(i)   **Magnetic field induction due to a current carrying straight conductor**

Consider a point P at a distance R from the straight conductor.

By symmetry all points at distance r will be on a circle of radius R.

Using Ampere's circuital law for P

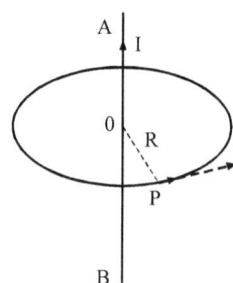

$$\oint \vec{B} \cdot \vec{dl} = \mu_0 I$$

$$B \oint dl = \mu_0 I$$

$$B \, 2\pi R = \mu_0 I$$

$\Rightarrow$
$$\boxed{B = \frac{\mu_0 I}{2\pi R}}$$

(ii)   **Magnetic field inside an infinitely long current carrying solenoid**

If there is a long solenoid of length l as shown.

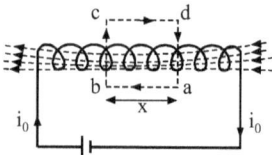

Experimentally it has been observed that magnetic field outside is very small compared with the field inside.

Applying Ampere's circuital law

$$\oint \vec{B} \cdot \vec{dl} = \mu_0 i$$

$$\oint \vec{B} \cdot \vec{dl} = \int_a^b \vec{B} \cdot \vec{dl} + \int_b^c \vec{B} \cdot \vec{dl} + \int_c^d \vec{B} \cdot \vec{dl} + \int_d^a \vec{B} \cdot \vec{dl}$$

here
$$\int_b^c \vec{B}.\,\overline{dl} = \int_d^a \vec{B}.\,\overline{dl} = 0$$

as both are perpendicular to field lines.

and
$$\int_c^d \vec{B}.\,\overline{dl} = 0 \text{ as } B = 0 \text{ outside the solenoid.}$$

so
$$\oint \vec{B}.\,\overline{dl} = \int_a^b \vec{B}.\,\overline{dl} = B\int_a^b dl = Bx \qquad \text{...(i)}$$

if there are n turns per unit length with i_0 current in each turn then within length of x, net current enclosed

$$i = nxi_0 \qquad \text{...(ii)}$$

form (i) & (ii)

$$Bx = \mu_0 \, nx \, i_0$$

or
$$\boxed{B = \mu_0 \, n \, i_0} \qquad \text{...(iii)}$$

if the solenoid is wrapped on a core of permeability μ_m then

$$B = \mu_m \, n \, i_0 = \mu_0 \, \mu_r \, n \, i_0 \qquad \text{...(iv)}$$

(iii) Magnetic field of induction due to a current carrying cylinder

Consider a cylinder of radius R with current I passing through it.

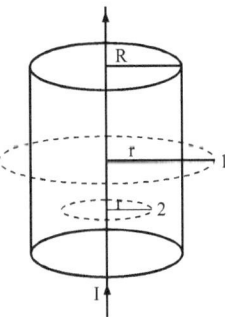

Magnetic field at any point at r distance form the axis of cylinder where (r > R) will be

$$\oint \vec{B}.\,\overline{dl} = B\oint dl = B2\pi r \qquad \text{...(i)}$$

as r > R current will pass through the circuit of radius r.

so $\qquad\qquad$ $B_2 \pi r = \mu_0 I$

$$B = \frac{\mu_0 I}{2\pi r}$$

...(ii)

It is same as for an infinite line current element.

When r < R

Consider a Gaussian surface of radius r inside the cylinder and current enclosed by the inner circle of radius r is given by

$$I' = \frac{\text{Current}}{\text{Area of the cylinder}} \times \text{area of the inner circle}$$

$$I' = \frac{I}{\pi R^2} \times \pi r^2 = \frac{I r^2}{R^2}$$

...(iii)

so $\qquad\qquad$ $B 2\pi r = \mu_0 I' = \mu_0 \dfrac{I r^2}{R^2}$

so $\qquad\qquad$

$$B = \frac{\mu_0 I r}{2\pi R^2} \quad \text{for } r < R$$

...(iv)

Inside a hollow cylinder

When there is a hollow cylinder, then whole current will exist only on the surface of the cylinder and inside current will become zero, so I = 0

hence $\qquad\qquad$ $B 2\pi r = 0$

or $\qquad\qquad$ $\boxed{B = 0}$

Time varying fields

When any charge is at rest it produces electrostatic fields and when charge moves with constant velocity, magnetic field is produces . Here electric and magnetic fields are independent of each other in the time invariant fields.

Whenever any charge moves with an acceleration, it produces a time varying electric field which then produces changing magnetic field. In the same manner a time varying magnetic field produce a changing electric field. Hence oscillating electric field produces an oscillating magnetic field and oscillating magnetic field produces oscillating electric field resulting in a wave like electric and magnetic fields known as electromagnetic wave, which can transport energy over a very long distances. Michael Faraday proposed the basis of electromagnetic field concept and gave postulates for electromagnetic induction to relate time varying fields.

Electromagnetic Induction

Faraday in 1831, found that when there is a change in magnetic flux passing through any circuit, an e.m.f. is generated in the circuit. A current flows in the circuit known as induced current and emf as induced emf. This lasts till the change in flux exists. As soon as the change stops, no emf or current is induced. This whole phenomenon is called electromagnetic induction.

Faraday's Laws of Electromagnetic Induction

1ˢᵗ Law : When number of magnetic lines of force attached to a closed circuit in changed, an emf is induced in the coil which lasts up to the change of field lines.

2nd law : The magnitude of induced emf is directly proportional to the rate of change of magnetic flux associated with the coil.

If $\Delta\phi$ is the change in magnetic flux in Δt time interval than the induced emf.

$$e = \frac{\Delta\phi}{\Delta t}$$

as $\Delta t \to 0$

$$e = \lim_{\Delta t \to 0} \frac{\Delta\phi}{\Delta t}$$

$$\boxed{e = \frac{d\phi}{dt}} \text{ volts}$$

if there are N turns in the coil, then the emf induced in the complete coil

$$e = N\frac{d\phi}{dt} \text{ or } \frac{d(N\phi)}{dt}$$

here $N\phi$ is called number of magnetic flux linkages. Its unit is weber turns.

3ʳᵈ law : The direction of induced emf is such that it always opposes its cause of generation. It is called Lenz's law.

Hence $\qquad\qquad e = -N\frac{d\phi}{dt}$

accordingly direction of induced emf can be found using flemming right hand rule.

Integral Forms

From Faraday's law of electromagnetic induction

$$e = -\frac{d\phi}{dt} \qquad \text{...(i)}$$

and magnetic flux ϕ through a closed surface is

$$\phi = \oint_s \vec{B} \cdot \vec{ds} \qquad \text{...(2)}$$

If the electric field induced is $\vec{E}$, then induced emf around the boundaries of closed surface is

$$e = \oint_c \vec{E} \cdot \vec{dl} \qquad \text{...(3)}$$

from (i), (ii) and (iii)

$$\oint_c \vec{E} \cdot \vec{dl} = -\frac{d}{dt} \oint_s \vec{B} \cdot \vec{ds}$$

or

$$\boxed{\oint_c \vec{E} \cdot \vec{dl} = -\oint_s \frac{\partial \vec{B}}{\partial t} \cdot \vec{ds}} \qquad \text{...(4)}$$

The equation (4) is known as the integral form of Faraday law of electromagnetic induction.

Differential form

From stoke's law we know that

$$\oint_c \vec{E} \cdot \vec{dl} = \oint_s \left(\vec{\nabla} \times \vec{E} \right) \cdot \vec{ds} \qquad \text{...(5)}$$

From (4) and (5)

$$\oint_s \left(\vec{\nabla} \times \vec{E} \right) \cdot \vec{ds} = -\oint_s \frac{\partial \vec{B}}{\partial t} \cdot \vec{ds}$$

or

$$\oint_s \left[\left(\vec{\nabla} \times \vec{E} \right) + \frac{\partial \vec{B}}{\partial t} \right] \cdot \vec{ds} = 0$$

or

$$\boxed{\vec{\nabla} \times \vec{E} = -\frac{\partial \vec{B}}{\partial t}} \qquad \text{...(6)}$$

Equation (6) represents the differential form of Faraday's law of electromagnetic induction.

SOLVED EXAMPLES

Q.1 : A parallel plate capacitor has a separation of 4 mm and potential difference of 200 volt between its plates. The capacitor is placed in a uniform magnetic field B . An electron projected vertically upward parallel to the plates with a velocity of 10^6 m/s, passes between the plates undeflected. Find the magnitude and direction of the magnetic field B between the plates. [IIT 81]

Sol.:

Suppose V is the potential difference between two plates A and B then

$$E = \frac{V}{d} = \frac{200V}{4 \times 10^{-3} \, m} = 5 \times 10^4 \text{ V/m}$$

and force due to this electric field on e^-

$$F = eE \qquad \qquad ...(i)$$

and force on the e^- due to presence of magnetic field

$$F = evB \qquad \qquad ...(ii)$$

so from (i) and (ii)

$$eE = evB$$

$$B = \frac{E}{v} = \frac{5 \times 10^4}{10^6} = 5 \times 10^{-2} \text{ wb/m}^2$$

Q.2 : An electrons is moving in a magnetic field of intensity 10^{-2} wb/m^2 with a velocity of 10^7 m/s in a circular path of radius 0.57 cm. Find out the specific charge of electron.

Sol.: When particle moves in circular path, necessary centripetal force $\dfrac{\left(mv^2 \right)}{r}$ is maintained with lorentz force due to magnetic field i.e. qvB.

So for electron

$$evB = \frac{mv^2}{r} \Rightarrow \frac{e}{m} = \frac{v}{Br}$$

$$\therefore \qquad \frac{e}{m} = \frac{10^7}{10^{-2} \times (0.57 \times 10^{-2})} = 1.76 \times 10^{11} \text{ C/kg}$$

Q.3 : An electron after being accelerated through a P. D. of 10^4 V enters a uniform magnetic field of 0.04 T perpendicular to its direction of motion. Calculate the radius of curvature of its trajectory.

Sol.: When electron is accelerated with potential V then

$$\frac{1}{2}mv^2 = eV \qquad\qquad\qquad ...(i)$$

and in magnetic field B, electron takes circular path and centripetal force

$$\frac{mv^2}{r} = evB \qquad\qquad\qquad ...(ii)$$

from (i) and (ii)

$$r = \frac{1}{B}\sqrt{\frac{2mV}{e}}$$

$$= \frac{1}{0.04}\sqrt{\frac{2 \times 9.1 \times 10^{-31} \times 10^4}{1.6 \times 10^{-19}}}$$

$$= 84.3 \times 10^{-4} \text{ m}$$

$$= 8.43 \text{ mm}$$

Q.4 : In cylindrical coordinates, $B = \left(\frac{4}{r}\right)\hat{\phi}$ T. Determine the magnetic flux ϕ crossing the plane surface given by $0.5 \leq r \leq 2.5$m and $0 \leq z \leq 2.0$m.

Sol.: We know that flux

$$\phi = \oint \vec{B} \cdot \overrightarrow{ds}$$

$$= \int_0^2 \int_{0.5}^{2.5} \left(\frac{4}{r}\right) \hat{a}_\rho \, dr \, dz \, \hat{\phi}$$

$$= 8 \ln \frac{2.5}{0.5} = 12.88 \text{ wb}$$

Q.5 : What is the magnitude of force on a wire of length 0.02 m placed inside a solenoid near its centre making an angle of 30° with its axis? The wire carries a current of 6A and the magnetic field due to solenoid is 0-25 T.

Sol.: The force on any conductor of length l is given by

$$F = I\,l\,B\sin\theta$$

$$= 6 \times 0.02 \times 0.25 \times 0.5$$

$$= 0.015 \text{ N}.$$

Q.6 : A horizontal overhead power lines carries a current of 90A from East to West. Compute the magnetic field generated at a distance of 1.5 m below the line.

Sol.: The magnitude of magnetic field due to a long, straight conductor is given by

$$B = \frac{\mu_0 I}{2\pi R}$$

here

$$\frac{\mu_0}{2\pi} = 2 \times 10^{-7} \text{ N/A}^2$$

$$= \left(2 \times 10^{-7}\right) \times \frac{90}{1.5}$$

$$= 1.2 \times 10^{-5} \text{ NA}^{-1} \text{ m}^{-1}$$

Q.7 : A differential current element with 10 Amp current and length 2×10^{-3} m is located at (2, 0, 0). Calculate the magnetic field $\vec{B}$ due to this element at (0, 0, 2).

Sol.: Here

$$I = 10 \text{ amp}$$

$$dl = 2 \times 10^{-3} \text{ m}$$

and

$$\vec{dl} = 2 \times 10^{-3}\,\hat{i} \text{ m}$$

30

$$I \cdot dl = 2 \times 10^{-2}\,\hat{i}$$

the distance between point and current element

$$\vec{r} = (0-2)\hat{i} + (0-0)\hat{j} + (2-0)\hat{k}$$

$$= -2\hat{i} + 2\hat{k}$$

$$|\vec{r}| = \sqrt{2^2 + 2^2} = \sqrt{8} = 2\sqrt{2}$$

and
$$\hat{r} = \frac{-2\hat{i} + 2\hat{k}}{\sqrt{8}} = \frac{2(-\hat{i} + \hat{k})}{2\sqrt{2}}$$

$$= \frac{-\hat{i} + \hat{k}}{\sqrt{2}}$$

from Biot Savart law

$$\vec{B} = \frac{\mu_0}{4\pi} \frac{I \, \vec{dl} \times \hat{r}}{r^2}$$

$$= \frac{4\pi \times 10^{-7}}{4\pi} \times \frac{0.02 \, \hat{i}}{\left(\sqrt{2}\right)^2} \times \frac{2(-\hat{i} + \hat{k})}{\sqrt{2}}$$

$$= -\frac{0.02 \times 10^{-7}}{\sqrt{2}} \, \hat{j}$$

$$= -\frac{2 \times 10^{-9}}{\sqrt{2}} \, \hat{j} = -\sqrt{2} \times 10^{-9} \, \hat{j}$$

$$= -1.414 \times 10^{-9} \, \hat{j} \quad \text{weber/m}$$

Q.8 : A helium nucleus is completing one round of a circle of radius 0.8 m in 4 seconds. Show that the magnetic field at the centre of the circle is 0.5 × 10⁻¹⁹ μ_0 T.

Sol.: As we know that helium nucleus has a charge of + 2e hence it is considered as a circle of radius r meter equivalent to a current loop and the centre, the magnetic field

$$B = \frac{\mu_0 I}{2r} \, T$$

if one revolution takes t sec then the current is

$$I = \frac{2e}{t} \, \text{amp}$$

so
$$B = \frac{\mu_0 2e}{2rt} = \frac{\mu_0 e}{rt}$$

$$= \frac{\mu_0 \times 1.6 \times 10^{-19}}{0.8 \times 4}$$

$$= 0.5 \times 10^{-19} \, \mu_0 \, T.$$

Q.9 : The magnetic flux threading a coil changes form 12×10^{-3} wb to 6×10^{-3} wb in 0.01 s. Calculate the induced emf.

Sol.: According to Faraday's law, the induced emf is

$$e = -\frac{\Delta(N \phi)}{\Delta t}$$

$$= -\frac{(6 \times 10^{-3}) - (12 \times 10^{-3})}{0.01}$$

$$= 0.6 \text{ wb/s}$$

$$= 0.6 \text{ V}$$

Q.10 : A coil having 100 turns and area 0.20 m² is placed normally in a magnetic field . the field changes from 0.20 wb/m² to 0.60 wb/m² uniformly over a period of 0.01 s. Calculate the emf induced in the coil.

Sol.: For each coil placed perpendicular to a magnetic field B is given by

$$\phi = BA$$

and change in flux due to change in B is

$$\Delta\phi = (\Delta B) \, A$$

$$= (0.60 - 0.20) \text{ wb} / \text{m}^2 \times 0.20\text{m}^2$$

$$= 0.08 \text{ wb}$$

By Faraday law, the magnitude of the induced emf is

$$|e| = N \frac{\Delta\phi}{\Delta t}$$

$$= \frac{100 \times 0.08}{0.01}$$

$$- 800 \text{ V}$$

Q.11 : When a magnetic flux lines changes from 5.5 × 10⁻⁴ to 5 × 10⁻⁵ in 0.1 sec through a coil of resistance 10 ohm with 1000 times. Find the electromotive force and the charge flowing through the coil.

Sol.: Here change in flux

$$\Delta\phi = (5 \times 10^{-5}) - (5.5 \times 10^{-4})$$

$$= -50 \times 10^{-5} \text{ wb}$$

hence induced emf

$$e = -N \frac{\Delta\phi}{\Delta t}$$

$$= -1000 \times \frac{\left(-50 \times 10^{-5}\right)}{0.1}$$

$$= 5V$$

the developed current $I = \dfrac{e}{R}$

$$= \frac{5V}{10\Omega} = 0.5 \text{ Amp}.$$

so the charge passed through the coil is

$$q = I \times \Delta t = 0.5 \times 0.1 = 0.05 \text{ C}$$

Q.12 : A current distribution gives rise to the vector potential
$\vec{A} = x^2 y \hat{i} + y^2 \times \hat{j} - 4xyz \hat{k}$ **wb/m. Calculate (i)** $\vec{B}$ **at (–1, 2, 5) (ii) magnetic flux through the surfaces defined by z = 1, $0 \le x \le 1$, $-1 \le y \le 4$.**   **[RU 2001]**

Sol.: As curl of the vector potential will give the magnetic field intensity

$$B (-1, 2, 5) = \text{curl } \vec{A}$$

$$= \begin{vmatrix} \hat{i} & \hat{j} & \hat{k} \\ \dfrac{\partial}{\partial x} & \dfrac{\partial}{\partial y} & \dfrac{\partial}{\partial z} \\ x^2 y & y^2 x & -4xyz \end{vmatrix}$$

$$= [-4xz]\hat{i} - [-4yz]\hat{j} + \left(y^2 - x^2\right)\hat{k}$$

$$= 20\,\hat{i} + 40\hat{j} + 3\hat{k} \text{ wb/m}^2$$

and flux $\phi = \oint \vec{B}.\,\overline{ds}$

where $\vec{B} = -4xz\,\hat{i} + 4yz\,\hat{j} + \left(y^2 - x^2\right)k^2$

and $\qquad \overrightarrow{ds} = dx\ dy\ \hat{k}$

so the flux $\qquad \phi = \oint \overrightarrow{B} . \overrightarrow{ds}$

$$= \int_{-1}^{4} \int_{0}^{1} \left[-4xz\,i + 4yz\,\hat{j} + \left(y^2 - x^2\right)\hat{k} \right] dx\ dy\ \hat{k}$$

$$= \int_{-1}^{4} \int_{0}^{1} \left(y^2 - x^2\right) dx\ dy = \int_{-1}^{4} \left[y^2 \left(-\frac{1}{3}\right) \right] dy$$

$$= \left[\frac{y^3}{3} - \frac{y}{3} \right]_{-1}^{4} = \frac{60}{3} = 20\ wb$$

Q.13 : Copper has 8×10^{28} free conduction electrons/m³. A copper wire of length 2.0 m and cross sectional area 8×10^{-6} m² carrying a current and lying perpendicular to a magnetic field of 5×10^{-3} T experiences a force of 8×10^{-2} N. Calculate the drift velocity of free electrons in the wire.

Sol.: When free electrons move with velocity v_d in the wire then the current is

$$I = ne\ A\ v_d \qquad\qquad ...(i)$$

and force on the conductor

$$F = I\ Bl \qquad\qquad ...(ii)$$

so $\qquad I = \dfrac{8 \times 10^{-2}\,N}{\left(5 \times 10^{-3}\right) \times 2m}$

$$= 8A$$

and $\qquad v_d = \dfrac{I}{neA} = \dfrac{8A}{8 \times 10^{28} \times 1.6 \times 10^{-19} \times 8 \times 10^{-6}}$

$$= 0.78 \times 10^{-4}\ m/s$$

Q.17 : A current of 20 Amp flows in downward direction in a long straight vertical wire and magnetic flux density in horizontal direction is 2×10^{-5} T. what is the distance of the neutral point form the wire?

Sol.: Neutral point will be where horizontal magnetic field will become equal to the magnetic field produced by current carrying conductor.

So
$$B_H = \frac{\mu_0 I}{2\pi r}$$

$$2 \times 10^{-5} = \frac{4\pi \times 10^{-7} \times 20}{2\pi \times r}$$

$$r = 0.2 \text{ m}$$

Q.18 : A circular coil is placed in uniform magnetic field of 0.10 T normal to the plane of the coil. If the current is 5.0 A in the coil, Find (a) total torque on the coil (b) total force on the coil (c) average force on each electron due to magnitude field (The coil in made of copper wire of cross section 10^{-5} m^2 and free electron density in copper is 10^{29} m^{-3}).

Sol.: We know that the torque on a current carrying coil of area A is

$$\tau = NIAB \sin \theta$$

(a)   here $\theta = 0$ i.e. $\sin \theta = 0$ so $\tau = 0$

(b)   the net forces will always zero.

(c)   and magnitude of force on a free electron

$$F = ev_d B$$

$$= \frac{IB}{nA} \qquad\qquad (\text{as} \quad I = neAv_d)$$

$$= \frac{5 \times 0.10}{10^{29} \times 10^{-5}} = 5 \times 10^{-25} \text{ N}$$

Q.19 : In the figure two current carrying wires are A and B. Find the magnitude and directions of the magnetic field at points 1, 2 and 3.

Sol.:

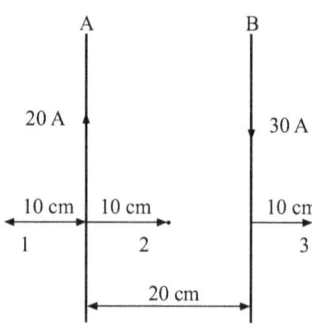

We know that magnetic field at R distance form straight current carrying wire

$$B = \frac{\mu_0 I}{2\pi R}$$

$$= \left(2\times10^{-7}\right)\frac{I}{R} \ NA^{-1} \ m^{-1}$$

Due to current in wire A at point 1 field will perpendicular upward and at 2 and 3 downwards and similarly due to current in wire B, 1 and 2 will be downward and 3 will be upward.

Hence points 1 and 3 are in opposite fields whereas point 3 will be always in same direction.

So resultant field at 1

$$B = B_1 - B_2$$

$$= \left(2\times10^{-7}\right)\frac{20}{0.1} - \left(2\times10^{-7}\right)\times\frac{30}{0.3}$$

$$= 2\times10^{-5} \ NA^{-1}m^{-1} \ \text{ perpendicular and upward to page.}$$

at 2

$$B = B_1 + B_2$$

$$= \left(2\times10^{-7}\right)\left[\frac{20}{0.1} + \frac{30}{0.3}\right]$$

$$= 1\times10^{-4} \ NA^{-1}m^{-1} \ \text{ perpendicular and downward to page.}$$

at 3

$$B = B_2 - B_1$$

$$= \left(2\times10^{-7}\right)\left[\frac{30}{0.1} - \frac{20}{0.3}\right]$$

$$= 4.7\times10^{-5} \ NA^{-1}m^{-1} \text{perpendicular and upward to the page.}$$

Q.20 : 8.0 cm length of a conductor is placed parallel to 2m length of a conductor at a distance of 2.0 cm. The conductors carry currents of 2 and 5 Amp respectively in opposite direction. Find the total force exerted on the long conductor.

Sol.: We know the magnetic field near to any straight conductor

$$B = \frac{\mu_0 I_1}{2\pi R}$$

So the force experienced by the short conductor carrying a current I_2

$$F = I_2 \ Bl$$

$$= \frac{\mu_0}{2\pi} \frac{I_1 I_2 l}{R}$$

so

$$F = \left(2 \times 10^{-7}\right) \times \frac{5 \times 2 \times 8 \times 10^{-2}}{2 \times 10^{-2}}$$

$$= 8 \times 10^{-6} \, N$$

according to Newton's third law, the long conductor will also experience an equal repulsive force 8×10^{-6} N due to the small conductor.

Q.21 : An air-cored solenoid of length 50 cm and area of cross-section 28 cm² has 200 turns and carries a current of 5A. On switching off, the current decreases to zero within a time interval of 2 ms. Find the average induced emf across the ends of the switch.

Sol.: As we know magnetic field

$$B = \mu_0 n \ I \qquad\qquad\qquad (n = \text{No. of turns/length})$$

$$= \left(4\pi \times 10^{-7}\right) \times \left(\frac{200}{50 \times 10^{-2}}\right) \times 5$$

$$= 25 \times 10^{-4} \ T$$

when switch off, flux reduces to zero, so change in flux.

$$\Delta\phi = 0 - NBA$$

$$= -200 \times \left(25 \times 10^{-4}\right) \times \left(28 \times 10^{-4}\right)$$

$$= -14 \times 10^{-4} \ wb$$

and induced emf

$$e = -\frac{\Delta\phi}{\Delta t} = \frac{14 \times 10^{-7}}{2 \times 10^{-3}} = 0.7 \ V$$

SUMMARY

- When any electric charge moves in a combined effect of electric and magnetic field then total force experienced by the charge

$$\vec{F} = q\left[\vec{E} + \vec{v} \times \vec{B}\right]$$

- If a current carrying conductor of length is placed in any magnetic field then the force experienced by that length of the conductor

$$F = I \vec{l} \times \vec{B}$$

$$= I \, lb \sin \theta$$

- Magnetic flux

$$\phi_B = \oint_s \vec{B} . \overrightarrow{ds}$$

- Gauss law of magnetism

$$\oint_s \vec{B} . \overrightarrow{ds} = 0$$

- Divergence of magnetic field is given as

$$\vec{\nabla} . \vec{B} = 0$$

which shows that magnetic poles exist in pairs.

- According to Biot-Savart law, the magnetic field at a point near to any current carrying conductor is given by

$$d\vec{B} = \frac{\mu_0}{4\pi} \frac{I \, \overrightarrow{dl} \times \hat{r}}{r^2}$$

- Magnetic field at the centre of a current carrying loop of radius r.

$$B = \frac{\mu_0 I}{2r}$$

- Magnetic field at a point due to current carrying straight conductor

$$B = \frac{\mu_0}{4\pi} \frac{I}{R} \left(\sin \phi_1 + \sin \phi_2\right)$$

$$B = \frac{\mu_0 I}{2\pi R} \, NA^{-1}m^{-1}$$

- Force between two parallel current carrying conductors

$$F = \frac{\mu_0}{2\pi} \frac{I_1 I_2}{R} \, l$$

this is attractive when currents are in same direction and repulsive when currents are in opposite direction.

- Magnetic field at the axis of current carrying circular coil

$$B = \frac{\mu_0 \, NI \, a^2}{2\left(a^2 + x^2\right)^{3/2}} \, NA^{-1}m^{-1}$$

- Magnetic field of a solenoid

$$B = \mu_0 nI \quad \text{at the centre}$$

$$= \frac{\mu_0 nI}{2} \quad \text{at the ends.}$$

- According to Ampere's circuital law the line integral of magnetic field due to a closed current carrying loop

$$\oint_c \vec{B} \cdot \vec{dl} = \mu_0 I \quad \text{Integral form}$$

$$\vec{\nabla} \times \vec{B} = \mu_0 \vec{J} \quad \text{Differential form}$$

- Magnetic field due to a current carrying cylinder

$$B = \frac{\mu_0 I r}{2\pi R^2} \quad \text{for } r < R$$

$$= \frac{\mu_0 I}{2\pi r} \quad \text{for } r > R$$

$$= 0 \quad \text{for a hollow cylinder and } r < R.$$

- According to 1st law of electromagnetic induction, when ever there is a change in magnetic flux linked with any closed solenoid then there is emf produced known as induced emf.

- According to 2nd law of electromagnetic induction the induced emf is given by

$$e = \frac{\Delta\phi}{\Delta t}$$

- According to 3rd law of EMI, the direction of induced emf is such as to oppose its cause of production so

$$e = -\frac{\Delta\phi}{\Delta t}, \text{ this is known as Lenz's law.}$$

- $\oint_c \vec{E}.\,\vec{dl} = -\oint_s \frac{\partial\vec{B}}{\partial t} \cdot \vec{ds}$ integral form of Faraday electromagnetic law.

- $\vec{\nabla} \times \vec{E} = -\frac{\partial\vec{B}}{\partial t}$ Differential form of Faraday's electromagnetic law.

- **Right hand palm rule No. 2**

  If we stretch palm of right hand such that the stretched fingers points towards the magnetic field direction and thumb points to the direction of current the force on the conductor will be perpendicular to the palm.

  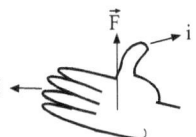

- **Flemming's left hand rule** –

  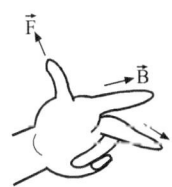

  If the thumb, middle finger and fore finger of the left hand are stretched in mutually perpendicular direction such that fore finger points in the direction of magnetic field $\vec{B}$ and middle finger towards the current direction then thumb will point in force $\vec{P}$ direction on the conductor.

- **Right hand palm rule no. 1**

  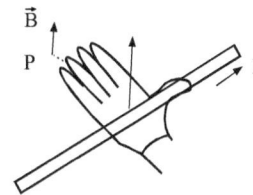

  If we stretch right hand palm such that thumb points in current direction and fingers towards the P point at which magnetic field direction is to be found, then the perpendicular to the palm will show the direction of $\vec{B}$ at P.

- **Right hand thumb rule –**

  It is used to find the direction of magnetic field due to a straight current carrying conductor. If we hold the straight conductor in our right hand in such away that thumb points in current direction, then the encircled fingers represents the magnetic field lines around the conductor.

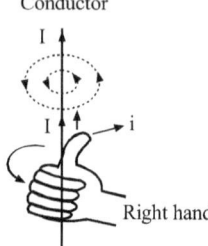
Conductor

Right hand

- **Maxwell's right hand screw rule**

Current I

Right hand

  If we takes the screw driver in our right hand such that it points to current direction and if we rotates it in such a manner that screw moves in the direction of current, then the direction of rotation of screw driver will tell the direction of magnetic field lines.

- **Direction of Induced current - Fleming's right hand rule–**

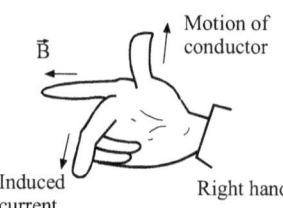
$\vec{B}$ Motion of conductor

Induced current   Right hand

  If we stretch the right hand thumb, fore finger and middle finger in mutually perpendicular direction in such a manner that fore finger is towards direction of magnetic field and thumb points in the direction of motion of conductor, then the middle finger will point in the direction of induced current.

EXERCISE

1. Describe the magnitude and direction of force acting on a charge moving in a magnetic field. When it is minimum and when it is maximum?

2. Derive an expression for the force experienced by a current-carrying straight conductor placed in a uniform magnetic field. State the rule to find the direction of this force.

3. Write Biot-Savart law for the magnetic field due to a current element, explaining the symbols.

4. Discuss analogies and differences between coulomb's law and Biot-Savart law.

5. A current is flowing through a thin, straight metallic conductor of infinite length. Find expression for the magnetic field at a distance from it.

6. Derive the relation for the force per unit length between two infinitely-long, parallel, straight conductors carrying current. Hence define one ampere.

7. Derive an expression for the magnetic field at a point on the axis of a circular coil carrying current, and hence at the centre of the coil.

8. Derive the expression for the magnetic field at the centre of a circular current carrying coil.

9. Deduce the expression for the magnetic field produced at the centre of a semicircular wire loop of radius R, carrying a current I.

10. Describe the magnetic field within a long, current carrying solenoid. Obtain expressions for the field within and at the ends of the solenoid.

11. State and explain Ampere's circuited law. Hence derive an expression for the magnetic field due to a solenoid.

12. Calculate, using Ampere's circuital law, the magnetic field due to infinitely-long current carrying conductor.

13. State and explain Faraday's laws of electromagnetic induction.

14. Show that lenz's law follows the principle of conservation of energy.

15. Give integral form of Faraday's laws of electromagnetic induction and convert these into differential forms.

16. Define magnetic flux. Write SI units of magnetic flux and magnetic flux density.

17. Write the formula for the force on a charge q moving with velocity $\vec{v}$ in a uniform magnetic field $\vec{B}$. What is the magnitude of this force? When this will be zero?

18. Prove that $\vec{\nabla}.\vec{B} = 0$ where $\vec{B}$ is the magnetic flux density.

19. What is vector potential? How the vector field can be calculated?

20. Prove that $\vec{\nabla} \times \vec{B} = \mu_0 \vec{J}$.

21. Deduce the differential form of Ampere's circuital law and hence prove that static magnetic field is not conservative.

22. Define electromagnetic wave. Can accelerated charged particle produce electromagnetic wave? give reasons.

23. An electron is moving vertically upward with a speed of 2×10^8 m/s. What will be the magnitude and direction of the force on the electron exerted by a horizontal magnetic field of 0.50 wb/m^2 directed towards west? What will be the acceleration of the electron?   [1.6×10^{-11}N north, 1.8×10^{19} m/s^2]

24. An electron moving with velocity 5×10^7 m/s enters a magnetic field of 1.0 wb/m^2 at an angle of 30 to the field. Calculate the force on the electron.

                                                         [4×10^{-12} N]

25. A 2-MeV proton is moving perpendicular to a uniform magnetic field of 2.5 T. Find the force on the proton. [Given mass of proton $= 1.65 \times 10^{-27}$ kg]

                                                  [Ans.: 7.88×10^{-12} N]

26. A 40 cm long wire carrying a current of 2.5A is placed perpendicular to a magnetic field of 8×10^{-3} wb/m^2. Find the force experienced by the wire.   [8×10^{-3} N]

27. A current of 5.0 A is flowing upward in a long vertical wire placed in a uniform horizontal north-ward magnetic field of 0.0207. How much forces and in what direction will the field exert on 0.06 m length of the wire.   [6×10^{-3} N, west]

28. A straights wire carries a current of 3A. Calculate the magnitude of the magnetic field at a point 10 cm away from the wire. Draw a diagram to show the direction of the magnetic field.   [6×10^{-6} T]

29. A circular loop of radius 5 cm carries a current of 0.5 amp. Calculate the magnitude of the magnetic field at its centre.   [6.28×10^{-6} N/A-m]

30. Calculate the force per unit length on a long straight wire carrying a current 4 amp due to a parallel wire carrying a current of 6 A. The distance between the wires is 3.0 cm.   [1.6×10^{-4} N/m]

31. Two parallel wires, each of length 2 m and carrying a current of 0.40 A in the same direction, are placed 0.40 m apart in air. Find the force per unit length on each wire.

                                                  [8×10^{-8} N/m attractive]

32. An Air-solenoid has 500 turn of wire in its 40 cm length. If the current in the wire be 1.0 A, find the magnetic field at the axis inside the solenoid.

$$[1.57 \times 10^{-3} \text{ NA}^{-1} \text{ m}^{-1}]$$

33. The magnetic field at the centre of a 50 cm long solenoid is 4×10^{-2} N/(A-m) when a current of 8 amp flows through it. What is the number of turns in the solenoid.

[1990]

34. A 0.5 m long solenoid has 500 turns and has a flux density of 2.52×10^{-3} T at its centre. Find the current in the solenoid.

[2.0 Amp.]

35. A test charge having charge 0.4 C is moving with a velocity of $\left(4\hat{i} - \hat{j} + 2\hat{k}\right)$ m/s through an electric field of intensity $10\hat{i} + 10\hat{k}$ and a magnetic field $2\hat{i} - 6\hat{j} - 6\hat{k}$. Determine the magnitude and direction of the lorentz force acting on the test charge.

[WBUT 2007]

36. If the vector potential $A = \left(x^2 + y^2 - z^2\right)\hat{j}$ at $\left(x_1 y_1 z\right)$. Find the magnetic field at (1, 1, 1).

[WBUT 2007]

ELECTROMAGNETIC THEORY

Concept of displacement current, Maxwell's field equations, Maxwell's wave equation and its solution for free space, Electromagnetic wave in a charge free conducting media, skin depth, physical significance of skin depth, Electromagnetic energy flow and Poynting vector.

As we know that magnetic field can be produced by a moving charge i.e. current in a conductor.

According to Ampere's circuital law, line integral of the magnetic field $\vec{B}$ arround any closed loop is equal to μ_0 times the net current threading through the area enclosed by the loop.

i.e.
$$\oint_C \vec{B} \cdot \overline{dl} = \mu_0 i$$

But when this equation is applied to charging and discharging of a capacitor, then according to Maxwell, it becomes inconsistent.

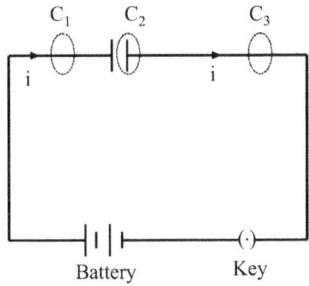

Battery   Key

Let us assume circular loops C_1, C_2 and C_3 around conductor before, in between and after the capacitor plates respectively.

If we now apply Ampere's circuital law on the above three loops then

$$\oint_{C_1} \vec{B} \cdot \overline{dl} = \oint_{C_3} \vec{B} \cdot \overline{dl} = \mu_0 i$$

But
$$\oint_{C_2} \vec{B} \cdot \overline{dl} = 0 \text{ [as there is no current in between capacitor plates.]}$$

so
$$\oint_{C_1} \vec{B} \cdot \overline{dl} = \oint_{C_3} \vec{B} \cdot \overline{dl} \neq \oint_{C_2} \vec{B} \cdot \overline{dl}$$

Which shows the inconsistency in the current flow.

Maxwell suggested that this contradiction arises due to assumed discontinuity in current between the plates. Maxwells suggested that during charging of the capacitor a changing electric field exists in between plates which is responsible for induced current in between plates. This lasts as long as electric field is changing i.e. capacitor is under the process of charging. This current corresponding to the changing electric field is called the displacement current. Hence during the process of capacitor charging, a conduction current i exists through connecting wires and a displacement current i_d in between capacitor plates. It makes a continuous current.

From the equation of continuity.

$$i_d = \varepsilon_0 \frac{d\phi_E}{dt}$$ where ε_0 is the permittivity constant.

so $$\oint \vec{B}.\,\vec{dl} = \mu_0 \left(i + i_d \right)$$

$$= \mu_0 \left[i + \varepsilon_0 \frac{d\phi_E}{dt} \right]$$

In case of ideal capacitor the amount of conduction current in the connecting wires is equal to the displacement current through the dielectric medium in between the walls of the capacitor.

hence $i_d = \dfrac{dq}{dt}$ where q is the induced charge on the capacitor plates.

$$= \frac{d}{dt} \left(A\varepsilon_0 E \right) \quad \left[\text{since } E = \frac{q}{\varepsilon_0\, A} \right]$$

or $$\boxed{i_d = A\, \varepsilon_0\, \frac{\partial E}{\partial t}}$$

The above equation shows a relation between displacement current and electric displacement.

The density of current i.e. displacement current density

$$\vec{J}_d = \frac{i_d}{A} = \varepsilon_0 \frac{\partial E}{\partial t}$$

$$= \frac{\partial \left(\varepsilon_0 E \right)}{\partial t} = \frac{\partial \vec{D}}{\partial t}$$

so
$$\boxed{\vec{J}_d = \frac{\partial \vec{D}}{\partial t}}$$

hence the rate of change of electric displacement vector w.r.t. time is equal to the displacement current density.

Hence experimentally, when a capacitor is connected to an attracting emf using a resistance, then due to time varying electric field, displacement current is setup between the dielectric medium of the capacitor so displacement current is the current setup in between capacitor plates due to time varying electric field in the dielectric medium.

As we know

$$E = \frac{q}{\varepsilon_0 A}$$

and
$$\frac{dE}{dt} = \frac{1}{\varepsilon_0 A} \frac{dq}{dt} = \frac{i}{\varepsilon_0 A}$$

so conduction current in connection wires

$$i = \varepsilon_0 A \frac{dE}{dt} \qquad \qquad ...(i)$$

By definition displacement current is

$$i_d - \varepsilon_0 \frac{d\phi_E}{dt}$$

$$= \varepsilon_0 A \frac{dE}{dt} \qquad \qquad ...(ii)$$

comparing (i) and (ii)

$$\boxed{i = i_d}$$

Hence displacement current in between plates is equal to the conduction current in the connecting wires. So both currents are continuous through the entire circuit if taken together but individually these are discontinuous.

The amount of magnetic effects are different from each current. A conduction current, due to its confinement to thin wire, produces much larger magnetic field in comparison to the displacement current which spreads through the entire surface area of plates of the capacitor.

Differences between conduction and displacement current

1. Conduction current obeys ohm's law as $i = \dfrac{V}{R}$ but displacement current does not obey ohm's law.

2. Conduction current density is represented by $\vec{J}_c = \sigma\vec{E}$ whereas displacement current density is given by $\vec{J}_d = \dfrac{\partial\vec{D}}{\partial t} = \varepsilon\dfrac{\partial\vec{E}}{\partial t}$.

3. Conduction current is the actual current whereas displacement current is the apparent current produced by time varying electric field.

Maxwell's equations

When charges move, electric and magnetic fields are associated with that motion which change in both space and time. There two fields are also interrelated. This phenomenon is known as electromagnetism which can be understood by the set of equations, known as Maxwell's equation. These are nothing but representations of the basic laws of electromagnetism.

Maxwell First Equation

This represents Gauss law of electrostatics according to which

$$\oint_s \vec{E}.\overrightarrow{ds} = \frac{q}{\varepsilon_0} = \frac{1}{\varepsilon_0}\int_V \rho\, dV$$

or $\qquad \oint_s \varepsilon_0\,\vec{E}.\overrightarrow{ds} = \int_V \rho\, dV$ (Integral form of maxwell's Ist equation.)

From Gauss divergences theorem $\oint_s \vec{D}.\overrightarrow{ds} = \int_V \left(\vec{\nabla}.\vec{D}\right) dV$

so $\qquad \int_V \left(\vec{\nabla}.\vec{D}\right) dV = \int_V \rho\, dV$

or $\boxed{\vec{\nabla}.\vec{D} = \rho}$ [Differential form of maxwell Ist equation]

Maxwell 2nd equation

It represents Gauss law of Magnetism which states that magnetic monopole does not exist, hence magnetic field induction across any closed surface is always zero.

$$\boxed{\oint_s \vec{B}.\overrightarrow{ds} = 0}$$ (Integral form of 2nd equation)

and from Gauss divergences theorem $\int_V \left(\vec{\nabla}.\vec{B}\right) dV = 0$

or $\boxed{\vec{\nabla}.\vec{B} = 0}$ (Differential from of maxwell's 2nd equation.)

Maxwell third equation

It represents Faraday's law of electromagnetic induction, according to which the curl of electric field vector is equal to the negative time rate of change of magnetic flux density.

from Faraday's law of electromagnetism, the induced emf.

$$e = -\frac{d\phi}{dt} \text{ where } \phi \text{ is the magnetic flux.} \quad \text{...(1)}$$

and induced emf through closed circuit bounding any surface S can be expressed in terms of electrostatic field $\vec{E}$ as $\qquad e = \oint_c \vec{E}.\overrightarrow{dl}$ ...(ii)

and from the definition of magnetic flux, $\qquad \phi = \oint_s \vec{B}.\overrightarrow{ds}$ ...(iii)

from (i) and (iii) $\qquad e = -\frac{d}{dt}\oint_s \vec{B}.\overrightarrow{ds}$ ...(iv)

from equation (ii) & (iv) $\qquad \oint_c \vec{E}.\overrightarrow{dl} = -\frac{d}{dt}\oint_s \vec{B}.\overrightarrow{ds}$ ...(v)

equation (v) represents integral from of Maxwells's IIIrd equation of electromagnetism.

Using stoke's law i.e.
$$\oint_c \vec{E} \cdot \overline{dl} = \oint_s (\vec{\nabla} \times \vec{E}) \cdot \overrightarrow{ds}$$

hence from (v)
$$\oint_s (\vec{\nabla} \times \vec{E}) \cdot \overrightarrow{ds} = -\frac{d}{dt} \oint_s \vec{B} \cdot \overrightarrow{ds}$$

or
$$\oint_s (\vec{\nabla} \times \vec{E}) \cdot \overrightarrow{ds} = -\oint \frac{\partial \vec{B}}{\partial t} \cdot \overrightarrow{ds}$$

hence
$$\boxed{\vec{\nabla} \times \vec{E} = -\frac{\partial \vec{B}}{\partial t}}$$
...(vi)

equation (vi) represents differential form of Maxwell's IIIrd equation of electromagnetism.

Maxwell's Fourth Equation

From the Ampere's circuital law
$$\oint_c \vec{B} \cdot \overline{dl} = \mu_0 (I_c + I_d)$$
...(i)

where I_c is the conduction current and I_d is the displacement current through the parallel plates of a capacitor due to time varying electric fields.

using
$$\vec{B} = \mu_0 \vec{H}$$

$$\oint_c \vec{H} \cdot \overline{dl} = I_c + I_d$$
...(ii)

Now from the definition of conduction current
$$I_c = \oint_s \vec{J}_c \cdot \overrightarrow{ds}$$
...(iii)

and displacement current
$$I_d = \oint_s \vec{J}_d \cdot \overrightarrow{ds}$$
...(iv)

so using (iii) and (iv) in (ii)

$$\oint_s \vec{H} \cdot \overline{dl} = \oint_s (\vec{J}_c + \vec{J}_d) \cdot \overrightarrow{ds}$$
...(v)

Here
$$J_d = \frac{\partial \vec{D}}{\partial t}$$

using (v)
$$\oint_c \vec{H}.\vec{dl} = \oint_s \left(\vec{J}_c + \frac{\partial \vec{D}}{\partial t} \right).\vec{ds} \qquad ...(vi)$$

equation (vi) is the integral from of Maxwell's fourth equation.

using stoke's law
$$\oint_s \vec{H}.\vec{dl} = \oint_s \left(\vec{\nabla} \times \vec{H} \right).\vec{ds} \text{ in (v)}$$

$$\oint_s \left(\vec{\nabla} \times \vec{H} \right).\vec{ds} = \oint_s \left(\vec{J}_c + \vec{J}_d \right).\vec{ds}$$

Hence
$$\vec{\nabla} \times \vec{H} = \vec{J}_c + \vec{J}_d$$

$$\boxed{\vec{\nabla} \times \vec{H} = \vec{J}_c + \frac{\partial \vec{D}}{\partial t}} \qquad ...(vii)$$

equation (vii) represents the differential form of Maxwell's fourth equation of electromagnetism.

So maxwell's equation can be summarized as

Maxwell's equation	Differential form	Integral form	In C. G. S.
I	$\vec{\nabla}.\vec{D} = \rho$	$\oint_s \vec{D}.\vec{ds} = \int_V \rho\, dV$	$\vec{\nabla}.\vec{D} = 4\pi\rho$
II	$\vec{\nabla}.\vec{B} = 0$	$\oint_s \vec{B}.\vec{ds} = 0$	—
III	$\vec{\nabla} \times \vec{E} = -\frac{\partial \vec{B}}{\partial t}$	$\oint_c \vec{E}.\vec{dl} = -\frac{\partial}{\partial t} \oint_s \vec{B}.\vec{ds}$	$\vec{\nabla} \times \vec{E} = -\frac{1}{c}\frac{\partial \vec{B}}{\partial t}$
IV	$\vec{\nabla} \times \vec{H} = \vec{J}_c + \frac{\partial \vec{D}}{\partial t}$	$\oint_c \vec{H}.\vec{dl} = \oint_s \vec{J}_c.\vec{ds} + \oint_s \frac{\partial \vec{D}}{\partial t}.\vec{ds}$	$\vec{\nabla} \times \vec{H} = 4\pi\, \vec{J}_c + \frac{1}{c}\frac{\partial \vec{D}}{\partial t}$

Physical significance of maxwell's Ist equation

According to this total electric flux through any closed surface is $\dfrac{1}{\varepsilon_0}$ times the total charge enclosed by the closed surfaces, representing Gauss's law of electrostatics, As this does not depend on time, it is a steady state equation. Here for positive ρ, divergence of electric field is positive and for negative ρ divergence is negative. It indicates that ρ is scalar quantity.

Physical significance of 2nd equation

It represents Gauss law of magnetostatic as $\vec{\nabla}.\vec{B} = 0$ resulting that isolated magnetic poles or magnetic monopoles cannot exist as they appear only in pairs and there is no source or sink for magnetic lines of forces. It is also independent of time i.e. steady state equation.

Physical significance of 3rd equation

It shows that with time varying magnetic flux, electric field is produced in accordance with Faraday is law of electromagnetic induction. This is a time dependent equation.

Physical significance of 4th equation

This is a time dependent equation which represents the modified differential form of Ampere's circital law according to which magnetic field is produced due to combined effect of conduction current density and displacement current density.

Maxwell's equations for different mediums

For free space (Vacuum)	For static fields (time variants)	For good conductors	For dielectrics
Here $\rho = 0$ $\vec{J}_c = \sigma\vec{E} = 0$	Here $\dfrac{\partial\vec{B}}{\partial t}=0, \dfrac{\partial\vec{D}}{\partial t} = 0$	Here $\rho = 0$ $J_d = 0$	Here $\sigma = 0$, $\rho = 0$ so $\vec{J}_c = 0$
$\vec{\nabla}.\vec{E} = 0$	$\vec{\nabla}.\vec{D} = \rho$	$\vec{\nabla}.\vec{D} = 0$	$\vec{\nabla}.\vec{D} = 0$
$\vec{\nabla}.\vec{H} = 0$	$\vec{\nabla}.\vec{B} = 0$	$\vec{\nabla}.\vec{B} = 0$	$\vec{\nabla}.\vec{B} = 0$
$\vec{\nabla}\times\vec{E} = -\mu_0\dfrac{\partial\vec{H}}{\partial t} = -\dfrac{\partial\vec{B}}{\partial t}$	$\vec{\nabla}\times\vec{E} = 0$	$\vec{\nabla}\times\vec{E} = -\dfrac{\partial\vec{B}}{\partial t}$	$\vec{\nabla}\times\vec{E} = -\dfrac{\partial\vec{B}}{\partial t}$
$\vec{\nabla}\times\vec{H} = \varepsilon_0\dfrac{\partial\vec{E}}{\partial t} = \dfrac{\partial\vec{D}}{\partial t}$	$\vec{\nabla}\times\vec{H} = \vec{J}_c$	$\vec{\nabla}\times\vec{H} = \vec{J}_c$	$\vec{\nabla}\times\vec{H} = \dfrac{\partial\vec{D}}{\partial t}$

Maxwell's equation in free space and its solution

For free space $\rho = 0$ i.e. $\vec{J}_c = 0$, $\sigma = 0$

Maxwell's equations becomes

$$\boxed{\vec{\nabla} \cdot \vec{D} = 0} ...(a) \quad \boxed{\vec{\nabla} \cdot \vec{B} = 0} ...(b) \quad \boxed{\vec{\nabla} \times \vec{E} = -\frac{\partial \vec{B}}{\partial t}} ...(c)$$

$$\vec{\nabla} \times \vec{H} = \frac{\partial \vec{D}}{\partial t} \quad \text{or} \quad \vec{\nabla} \times \frac{\vec{B}}{\mu_0} = \frac{\partial \vec{D}}{\partial t} \quad \text{or} \quad \boxed{\vec{\nabla} \times \vec{B} = \varepsilon_0 \, \mu_0 \, \frac{\partial \vec{E}}{\partial t}} ...(d)$$

taking curl of equation (c)

$$\vec{\nabla} \times \left(\vec{\nabla} \times \vec{E}\right) = \vec{\nabla} \times \left(-\frac{\partial \vec{B}}{\partial t}\right) = -\frac{\partial}{\partial t}\left(\vec{\nabla} \times \vec{B}\right)$$

or $\qquad \vec{\nabla}\left(\vec{\nabla} \cdot \vec{E}\right) - \nabla^2 \vec{E} = -\mu_0 \, \varepsilon_0 \, \frac{\partial}{\partial t}\left(\frac{\partial \vec{E}}{\partial t}\right)$ using (d)

for charge free region $\quad \vec{\nabla} \cdot \vec{D} = 0$ i.e. $\vec{\nabla} \cdot \vec{E} = 0$

so $\qquad -\nabla^2 \vec{E} = -\mu_0 \, \varepsilon_0 \, \frac{\partial^2 \vec{E}}{\partial t^2}$

or $\qquad \boxed{\nabla^2 \vec{E} = \mu_0 \, \varepsilon_0 \, \frac{\partial^2 \vec{E}}{\partial t^2}} \qquad\qquad ...(e)$

equation (e) is the maxwell wave equation for electric field in free space.

Now taking curl of equation (d)

$$\vec{\nabla} \times \vec{\nabla} \times \vec{B} = \mu_0 \, \varepsilon_0 \, \frac{\partial}{\partial t}\left(\vec{\nabla} \times \vec{E}\right)$$

or $\qquad \vec{\nabla}\left(\vec{\nabla} \cdot \vec{B}\right) - \nabla^2 \vec{B} = -\mu_0 \varepsilon_0 \, \frac{\partial \vec{B}}{\partial t} \qquad\qquad \left[\text{as } \vec{\nabla} \times \vec{E} = -\frac{\partial \vec{B}}{\partial t} \text{ from (c)}\right]$

But $\qquad \vec{\nabla}.\vec{B} = 0$

$$\nabla^2 \vec{B} = \mu_0\, \varepsilon_0 \frac{\partial^2 \vec{B}}{\partial t^2}$$

or $\qquad \boxed{\nabla^2\,\vec{H} = \mu_0\, \varepsilon_0 \frac{\partial^2 \vec{H}}{\partial t^2}}$ $\qquad$ ...(f)

equation (f) is the maxwell equation for magnetic field in free space.

comparing there equation with the general wave equation $\nabla^2\,\psi = \dfrac{1}{v^2}\dfrac{\partial^2 \psi}{\partial t^2}$

we get $\quad v^2 = \dfrac{1}{\mu_0\, \varepsilon_0}$ $\qquad$ or $\;\; v = \dfrac{1}{\sqrt{\mu_0\,\varepsilon_0}} = \dfrac{1}{\sqrt{4\pi \times 10^{-7} \times 8.854 \times 10^{-12}}}$

$\qquad\qquad = 2.998 \times 10^8 \simeq 3 \times 10^8$ m/s $\quad$ (equal to the speed of light in vacuum)

It clearly shows that $\vec{E}$ and $\vec{H}$ i.e. electromagnetic wave can propagate in free space with speed of light suggesting that the light is an electromagnetic wave.

Hence the solution of an electric wave for $\vec{E}$ can be written as

$$\vec{E}(\vec{r}, t) = \vec{E}_0\, e^{i(\vec{k}.\vec{r} - \omega t)} \qquad \text{...(i)}$$

and for magnetic field $\vec{H}$ it is

$$\vec{H}(\vec{r}, t) = \vec{H}_0 e^{i(\vec{k}.\vec{r} - \omega t)} \qquad \text{...(ii)}$$

here E_0 and H_0 are the maximum values and ω is the frequency of the wave. k represents the propagation vector for the wave.

Electromagnetic waves in a charge free conducting medium

for charge free conductive medium $\rho = 0$, $\qquad \vec{J}_c = \sigma \vec{E}$

$$\vec{\nabla}.\vec{E} = 0 \;\text{...(i)} \qquad \vec{\nabla}.\vec{H} = 0 \text{...(ii)} \qquad \vec{\nabla} \times \vec{E} = -\mu \frac{\partial \vec{H}}{\partial t} \text{...(iii)}$$

$$\vec{\nabla} \times \vec{H} = \vec{J}_c + \varepsilon \frac{\partial \vec{E}}{\partial t} = \sigma \vec{E} + \varepsilon \frac{\partial \vec{E}}{\partial t} \qquad \text{...(iv)}$$

taking curl of (iii) $\qquad \vec{\nabla} \times \vec{\nabla} \times \vec{E} = -\mu \vec{\nabla} \times \frac{\partial \vec{H}}{\partial t}$

or $\qquad \vec{\nabla}\left(\vec{\nabla}.\vec{E}\right) - \nabla^2 \vec{E} = -\mu \vec{\nabla} \times \frac{\partial \vec{H}}{\partial t}$

using (i) $\qquad -\nabla^2 \vec{E} = -\mu \vec{\nabla} \times \frac{\partial \vec{H}}{\partial t} \qquad \text{or} \quad \nabla^2 \vec{E} = \mu \frac{\partial}{\partial t}\left(\vec{\nabla} \times \vec{H}\right)$

using (iv) $\qquad \nabla^2 \vec{E} = \mu \frac{\partial}{\partial t}\left[\sigma \vec{E} + \varepsilon \frac{\partial \vec{E}}{\partial t}\right] = \mu\sigma \frac{\partial \vec{E}}{\partial t} + \mu\varepsilon \frac{\partial^2 \vec{E}}{\partial t^2}$

or $\qquad \boxed{\nabla^2 \vec{E} - \varepsilon\mu \frac{\partial^2 \vec{E}}{\partial t^2} - \mu\sigma \frac{\partial \vec{E}}{\partial t} = 0} \qquad \text{...(v)}$

Similarly taking curl of equation (iv)

$$\vec{\nabla} \times \left(\vec{\nabla} \times \vec{H}\right) = \vec{\nabla} \times \vec{J} + \vec{\nabla} \times \frac{\partial \vec{D}}{\partial t}$$

$$\vec{\nabla}\left(\vec{\nabla}.\vec{H}\right) - \nabla^2 \vec{H} = \sigma\left(\vec{\nabla} \times \vec{E}\right) - \varepsilon \frac{\partial}{\partial t}\left(\vec{\nabla} \times \vec{E}\right)$$

using (ii) $\quad -\nabla^2 \vec{H} = -\sigma\mu \frac{\partial H}{\partial t} - \varepsilon\mu \frac{\partial^2 H}{\partial t^2} \qquad \text{or} \qquad \nabla^2 \vec{H} = \sigma\mu \frac{\partial \vec{H}}{\partial t} + \varepsilon\mu \frac{\partial^2 \vec{H}}{\partial t^2}$

or $\qquad \boxed{\nabla^2 \vec{H} - \sigma\mu \frac{\partial H}{\partial t} - \varepsilon\mu \frac{\partial^2 \vec{H}}{\partial t^2} = 0} \qquad \text{...(vi)}$

These general equation (v) and (vi) represents general wave equations for a conducting medium with conductivity σ. The behaviour of electromagnetic fields in uniform and charge free conducting medium is characterized by these equations.

A conductive medium is also called lossy medium as the terms $\mu\sigma\dfrac{\partial\vec{E}}{\partial t}$ and $\mu\sigma\dfrac{\partial\vec{H}}{\partial t}$ in general wave equations represent electromagnetic field decay i.e. loss of energy. Hence there is always a loss of energy when any electromagnetic wave propagates through the medium.

In charge free non conducting dielectric medium $\rho = 0$ as well as $\sigma = 0$ making no energy loss that is why dielectric medium is called loss less medium.

Loss of energy i.e. attenuation (concept of skin depth)

As the general wave equation in a charge free and external current free medium is given by

$$\nabla^2\vec{E} - \varepsilon\mu\frac{\partial^2\vec{E}}{\partial t^2} - \sigma\mu\frac{\partial\vec{E}}{\partial t} = 0 \ldots\text{(i)} \quad \text{and} \quad \nabla^2\vec{H} - \varepsilon\mu\frac{\partial^2\vec{H}}{\partial t^2} - \sigma\mu\frac{\partial\vec{H}}{\partial t} = 0 \ldots\text{(ii)}$$

to solve these equation, consider the solution be

$$\vec{E} = \vec{E}_0\, e^{i(\vec{k}.\vec{r} - \omega t)} \qquad \ldots\text{(iii)} \quad \text{and} \quad \vec{H} = \vec{H}_0\, e^{i(\vec{k}.\vec{r} - \omega t)} \qquad \ldots\text{(iv)}$$

so

$$\frac{\partial\vec{E}}{\partial t} = -i\omega\,\vec{E}_0\, e^{i(\vec{k}.\vec{r} - \omega t)} = -i\omega E \qquad \ldots\text{(v)}$$

and

$$\frac{\partial^2\vec{E}}{\partial t^2} = (-i\omega)^2\, E_0\, e^{i(\vec{k}.\vec{r} - \omega t)} = \omega^2\vec{E} \qquad \ldots\text{(vi)}$$

$$\nabla^2\vec{E} = \frac{\partial^2\vec{E}}{\partial r^2} = i^2 k^2\,\vec{E} = -k^2\,\vec{E} \qquad \ldots\text{(vii)}$$

substituting (v), (vi) and (vii) in (i)

$$-k^2\,\vec{E} + \varepsilon\mu\omega^2\,\vec{E} + i\sigma\omega\mu\,\vec{E} = 0$$

$$\left[k^2 - \varepsilon\mu\omega^2 - i\sigma\mu\varepsilon\right]\vec{E} = 0$$

or $\quad k^2 = \varepsilon\mu\omega^2 + i\sigma\mu\omega \qquad$ or $\quad k^2 = \varepsilon\mu\omega^2\left[1 + \dfrac{i\sigma}{\varepsilon\omega}\right] \qquad \ldots\text{(viii)}$

hence $\vec{k}$ is the propagation constant which represent the dispersion relation for any electromagnetic wave in a lossy dielectric medium and provides information about the nature of propagation of electromagnetic wave inside any medium.

$\vec{k}$ being a complex quantity so $\quad k = A + iB$

$$k^2 = (A+iB)(A+iB) \quad = A^2 - B^2 + i\,2AB \qquad \text{...(ix)}$$

Comparing equation (viii) and (ix)

$$A^2 - B^2 = \varepsilon\mu\omega^2 \quad \text{and} \quad 2AB = \sigma\mu\omega$$

On solving these two equations

$$A = \omega\sqrt{\frac{\varepsilon\mu}{2}}\left[\left\{1+\left(\frac{\sigma}{\varepsilon\omega}\right)^2\right\}^{1/2} + 1\right]^{1/2} \qquad \text{...(x)}$$

$$B = \omega\sqrt{\frac{\varepsilon\mu}{2}}\left[\left\{1+\left(\frac{\sigma}{\varepsilon\omega}\right)^2\right\}^{1/2} - 1\right]^{1/2} \qquad \text{...(xi)}$$

Here the term $\dfrac{\sigma}{\varepsilon\omega}$ is known as dissipation factor because it is the ratio of conduction current density to displacement current density. A is called attenuation constant and B is called as phase constant. In any loss less medium, waves do not attenuate so A = 0.

Hence equations for $\vec{E}$ and $\vec{H}$ can be written as

$$\vec{E} = \vec{E}_0\, e^{-Br}\, e^{i(Ar-\omega t)} \quad \text{...(xii)} \quad \text{and} \quad \vec{H} = \vec{H}_0\, e^{-Br}\, e^{i(Ar-\omega t)} \quad \text{...(xiii)}$$

here e^{-Br} is called attenuation factor and $e^{i(Ar-\omega t)}$ is called phase factor. e^{-Br} shows exponential decrease in amplitude with increase in r. Here B is called the absorption coefficient and is a measure of attenuation. For a good conducting medium $\mu\varepsilon\dfrac{\partial^2 E}{\partial t^2}$ is negligible as displacement current does not exist experimentally. Hence general wave equation reduces to

$$\nabla^2 \vec{E} - \mu\sigma \frac{\partial \vec{E}}{\partial t} = 0 \qquad \text{...(i)}$$

and the general solution for above is $\vec{E} = \vec{E}_0\, e^{-Br}\, e^{i(Ar-\omega t)}$ ...(ii)

for a good conductor, when the frequency of the electromagnetic wave is not very high then $\sigma \gg \varepsilon\omega$ i.e. $\dfrac{\sigma}{\varepsilon\omega} \gg 1$.

and
$$A = B = \omega\sqrt{\frac{\mu\varepsilon}{2}} \times \frac{\sigma}{\varepsilon\omega} = \sqrt{\frac{\mu\sigma\varepsilon}{2}} = \frac{1}{\delta}\,(\text{assume})$$

so
$$\delta = \sqrt{\frac{2}{\mu\sigma\varepsilon}} \qquad \text{...(iii)}$$

Here general solution becomes

$$\vec{E} = \vec{E}_0\, e^{-\frac{r}{\delta}}\, e^{i\left(\frac{r}{\delta}-\omega t\right)} \qquad \left[\text{as } A = B = \frac{1}{\delta}\right] \quad \text{...(iv)}$$

Hence the amplitude of the electromagnetic wave becomes $\vec{E}_0 e^{-r/\delta}$. If $r = \delta$ then amplitude becomes $\vec{E}_0 e^{-1}$ or $\dfrac{\vec{E}_0}{e}$. Hence we can define distance $r = \delta$ as depth of penetration or skin depth, where the amplitude of electric field reduces to $1/e$ times of the amplitude of electric field at the surface (i.e. $r = 0$).

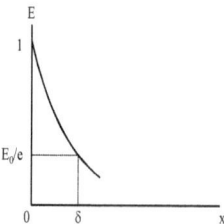

Attenuation of electric field in a conductor

Hence the skin depth is given as

$$\delta = \sqrt{\frac{2}{\mu\sigma\omega}} \qquad ...(v)$$

The skin depth decrease with increase in frequency of the electromagnetic wave and the conductivity of the medium.

Physical significance of skin depth

It shows that any electromagnetic wave with high frequency (ω) cannot propagate through the conducting media. The large value of skin depth signifies the less attenuation of electromagnetic waves in any medium.

Electromagnetic Energy density

We know that electrostatic field energy is $W_E = \frac{1}{2} \int_V \left(\vec{E}.\vec{D}\right)dV = \frac{\varepsilon_0}{2} \int E^2 \, dV \qquad ...(i)$

and magnetostatic field energy is $W_M = \frac{1}{2} \int \left(\vec{B}.\vec{H}\right)dV = \frac{1}{2\mu_0} \int B^2 dV \qquad ...(ii)$

So total energy stored in electromagnetic field

$$W_{EM} = \frac{1}{2} \int \left(\varepsilon_0 E^2 + \frac{B^2}{\mu_0} \right) dV \qquad ...(iii)$$

Hence electromagnetic energy density can be written as

$$\boxed{U_{EM} = \frac{1}{2} \left(\varepsilon_0 E^2 + \frac{B^2}{\mu_0} \right)} \qquad ...(iv)$$

Poynting vector and poynting theorem

When electromagnetic wave travels in space, it carries energy and energy density is always associated with electric fields and magnetic fields.

The rate of energy travelled through per unit area i.e. the amount of energy flowing through per unit area in the perpendicular direction to the incident energy per unit time is called poynting vector.

Mathematically poynting vector is represented as

$$\vec{P} = \vec{E} \times \vec{H} \left(= \frac{\vec{E} \times \vec{B}}{\mu} \right) \qquad \text{...(i)}$$

the direction of poynting vector is perpendicular to the plane containing $\vec{E}$ and $\vec{H}$. Poynting vector is also called as instantaneous energy flux density. Here rate of energy transfer $\vec{P}$ is perpendicular to both $\vec{E}$ and $\vec{H}$. Since it represents the rate of energy transfer per unit area, its unit is W/m².

Poynting theorem states that the net power flowing out of a given volume V is equal to the time rate of decrease of stored electromagnetic energy in that volume decreased by the conduction losses.

i.e. total power leaving the volume = rate of decrease of stored electromagnetic energy
– ohmic power dissipated due to motion of charge

Proof : The energy density carried by the electromagnetic wave can be calculated using Maxwell's equations

as $\quad$ div $\vec{D} = 0$ ...(i) $\quad$ div $\vec{B} = 0$ ...(ii) $\quad$ Curl $\vec{E} = -\dfrac{\partial \vec{B}}{\partial t}$ ...(iii)

$$\text{and Curl } \vec{H} = \vec{J} + \frac{\partial \vec{D}}{\partial t} \qquad \text{...(iv)}$$

taking scalar product of (iii) with H and (iv) with $\vec{E}$

i.e. $\qquad \vec{H} \text{ curl } \vec{E} = -\vec{H} \cdot \dfrac{\partial \vec{B}}{\partial t} \qquad \text{...(v)}$

and $\qquad \vec{E} \cdot \text{curl } \vec{H} = \vec{E}.\vec{J} + \vec{E}.\dfrac{\partial \vec{D}}{\partial t} \qquad \text{...(vi)}$

doing (vi) – (v) i.e. $\qquad \vec{H}.\text{curl } \vec{E} - \vec{E}. \text{curl } \vec{H} = -\vec{H}.\dfrac{\partial \vec{B}}{\partial t} - \vec{E}.\vec{J} - \vec{E}.\dfrac{\partial \vec{D}}{\partial t}$

$$= -\left[\vec{H}.\frac{\partial \vec{B}}{\partial t} + \vec{E}.\frac{\partial \vec{D}}{\partial t} \right] - \vec{E}.\vec{J}$$

as $\qquad$ div $\left(\vec{A} \times \vec{B}\right) = \vec{B}.\,\text{curl}\,\vec{A} - \vec{A}\,\text{curl}\,B$

so $\qquad$ div $\left(\vec{E} \times \vec{H}\right) = -\left[\vec{H}.\dfrac{\partial \vec{B}}{\partial t}+\vec{E}.\dfrac{\partial \vec{D}}{\partial t}\right]-\vec{E}.\vec{J}$ $\qquad$ ...(vii)

But $\qquad$ $\vec{B} = \mu\vec{H}$ and $\vec{D} = \varepsilon\vec{E}$

so $\qquad$ $\vec{H}.\dfrac{\partial \vec{B}}{\partial t} = \vec{H}.\dfrac{\partial}{\partial t}\left(\mu\vec{H}\right)=\dfrac{1}{2}\,\mu\,\dfrac{\partial}{\partial t}\left(H^2\right)$

$$= \dfrac{\partial}{\partial t}\left[\dfrac{1}{2}\,\vec{H}.\vec{B}\right]$$

and $\qquad$ $\vec{E}.\dfrac{\partial \vec{D}}{\partial t} = \vec{E}.\dfrac{\partial}{\partial t}\left(\varepsilon\vec{E}\right)=\dfrac{1}{2}\varepsilon\dfrac{\partial}{\partial t}(E)^2 =\dfrac{\partial}{\partial t}\left[\dfrac{1}{2}\,\vec{E}.\vec{D}\right]$

so from equation (vii) $\quad$ div $\left(\vec{E} \times\vec{H}\right) = -\dfrac{\partial}{\partial t}\left[\dfrac{1}{2}\left(\vec{H}.\vec{B} + \vec{E}.\vec{D}\right)\right]-\vec{E}.\vec{J}$

or $\qquad$ $\vec{E}.\vec{J} = -\dfrac{\partial}{\partial t}\left[\dfrac{1}{2}\left(\vec{H}.\vec{B} + \vec{E}.\vec{D}\right)\right]-\text{div}\left(\vec{E}\times\vec{H}\right)$ $\qquad$ ...(viii)

Integrating equation (viii) over a volume V enclosed by a surface S

$$\int_V \vec{E}.\vec{J}\,dV = -\int_V\left|\dfrac{\partial}{\partial t}\left\{\dfrac{1}{2}\left(\vec{H}.\vec{B}+\vec{E}.\vec{D}\right)\right\}\right|dV - \int_V \text{div}\left(\vec{E}\times\vec{H}\right)dV$$

or $\qquad$ $\displaystyle\int_V \vec{E}.\vec{J}\,dV = -\int_V\left[\dfrac{1}{2}\mu H^2 +\dfrac{1}{2}\varepsilon E^2\right]dV - \int_S\left(\vec{E}\times\vec{H}\right).ds$

as $\quad$ $\vec{B} =\mu\vec{H},$ $\quad$ $\vec{D} = \varepsilon\vec{E}$ and $\displaystyle\int_V \text{div}\left(\vec{E}\times\vec{H}\right)dV = \int_S\left(\vec{E}\times\vec{H}\right).ds$

or $\qquad$ $\displaystyle\int_V\left(\vec{E}.\vec{J}\right)dV = -\dfrac{\partial}{\partial t}\int_V\left[\dfrac{1}{2}\mu H^2 +\dfrac{1}{2}\varepsilon E^2\right]dV - \int_S\left(\vec{E}\times\vec{H}\right).ds$

or

$$\int_s \left(\vec{E} \times \vec{H}\right).ds = -\int_V \frac{\partial U_{em}}{\partial t} dV - \int_V \left(\vec{E}.\vec{J}\right) dV$$

or

$$\boxed{\int_s \vec{P}.ds = -\int_V \frac{\partial U_{em}}{\partial t} dV - \int_V \left(\vec{E}.\vec{J}\right) dV} \qquad \left(as\ \vec{P} = \vec{E} \times \vec{H}\right) \quad ...(ix)$$

i.e.   Total power leaving the volume = rate of decrease of stored e.m. energy -

ohmic power dissipated due to charge motion

This equation (ix) represents the poynting theorem according to which the net power flowing out of a given volume is equal to the rate of decrease of stored electromagnetic energy in that volume minus the conduction losses.

In equation (ix) $\int_s \vec{P}.ds$ represents the amount of electromagnetic energy crossing the closed surface per second or the rate of flow of outward energy through the surface S enclosing volume V i.e. it is poynting vector.

The term $\int_V \frac{\partial U_{em}}{\partial t} dV$ or $\frac{\partial}{\partial t} \int_V \left[\frac{1}{2} \mu H^2 + \frac{1}{2} \varepsilon E^2\right] dV$, the terms $\frac{1}{2}\mu H^2$ and $\frac{1}{2}\varepsilon E^2$ represent the energy stored in electric and magnetic fields respectively and their sum denotes the total energy stored in electromagnetic field. So total terms gives the rate of decrease of energy stored in volume V due to electric and magnetic fields.

$\int_V \left(\vec{E}.\vec{J}\right) dV$ gives the rate of energy transferred into the electromagnetic field.

This is also known as work-energy theorem. This is also called as the energy conservation law in electromagnetism.

SOLVED EXAMPLES

Example 1 : An air filled parallel plate capacitor has 11 pF capacitance. If the potential difference across the plates changes at a rate of 10^{12} V s^{-1}, what will be the value of displacement current?

Solution : We know that displacement current is

$$i_d = \varepsilon_0 \frac{d\phi_e}{dt} = \varepsilon_0 A \frac{dE}{dt} = \varepsilon_0 \frac{A}{d} \frac{dV}{dt}$$

$$= C \frac{dV}{dt} \qquad \left(as = C = \frac{\varepsilon_0 A}{d} \right)$$

$$= 11 \times 10^{-12} \times 10^{12}$$

$$= 11 \text{ Amp.}$$

Example 2 : Calculate the value of poynting vector at the surface of the sun if the power radiated by the sun is 4 × 10²⁶W and its radius is 7 × 10⁸m.

Solution : If S_{av} is the average poynting vector at surface of the sun

then $\qquad\qquad S_{av} \times 4\pi r^2 = P$

where P is the power radiated.

$$S_{av} = \frac{P}{4\pi r^2} = \frac{4 \times 10^{26}}{4 \times 3.14 \times \left(7 \times 10^8\right)^2}$$

$$= 6.499 \times 10^7 \ W/m^2$$

Example 3: A parallel plate capacitor is charged initially to 80μC. The area of capacitor plate is 500 cm² and plate separation is 3.00mm. The capacitor looses charge to the nearby medium at a rate of 1.5 × 10⁻⁸ cs⁻¹. What is the direction and magnitude of displacement current and magnetic field between the plates of the capacitor?

Solution: The rate of loosing charge will be equal to the conduction current within the plates.

and displacement current

$$i_d = \varepsilon_0 \frac{d\phi_e}{dt} = \varepsilon_0 \frac{d}{dt}(EA)$$

$$= \varepsilon_0 A \frac{d}{dt}\left(\frac{q}{\varepsilon_0 A}\right) \qquad \left(as \ E = \frac{q}{\varepsilon_0 A} \right)$$

$$\frac{dq}{dt} = 1.5 \times 10^{-8} \text{ cs}^{-1} = 1.5 \times 10^{-8} \text{A}$$

the direction i_d will be opposite to the direction of conduction current, hence total current

$$i' = i + i_d = 0$$

and using. Ampere's circuital law

$$\oint \vec{B} . \vec{dl} = \mu_0 i' = 0$$

so $B = 0$

So at all points between the plates magnetic field is zero.

Example 4 : In a plane electromagnetic wave, the electric field oscillates sinusoidally with a frequency of 2×10^{10} Hz and amplitude 48 Vm^{-1} (a) find the wavelength of the wave? (b) find the amplitude of the oscillating magnetic field? (c) calculates the total average energy density of the electromagnetic field of wave.   [AISSCE 90]

Solution: (a)   the wavelength of electromagnetic wave is

$$\lambda = \frac{c}{V} = \frac{3 \times 10^8 \text{ m/s}}{2 \times 10^{10} \text{ s}^{-1}} = 1.5 \times 10^{-2} \text{ m}$$

(b)   as $E_0 = CB_0$

$\Rightarrow$ $$B_0 = \frac{E_0}{C} = \frac{48 \text{ Vm}^{-1}}{3 \times 10^8 \text{ ms}^{-1}} = 1.6 \times 10^7 \text{ NA}^{-1}\text{m}^{-1}$$

(c)   The total average energy density of the electromagnetic field is equal to the maximum electric energy density $\left(\text{i.e. } \frac{1}{2} \varepsilon_0 E_0^2 \right)$ or maximum magnetic energy density $\left(\text{i.e. } \frac{1}{2} B_0^2 / \mu_0 \right)$.So

$$U = \frac{1}{2} \varepsilon_0 E_0^2$$

$$= \frac{1}{2} \times \left(8.85 \times 10^{-12} \right) \times (48)^2$$

$$= 1.0 \times 10^{-8} \text{ Nm}^{-2}$$

$$= 1.0 \times 10^{-8} \text{ Jm}^{-3}$$

Example 5 : Electromagnetic waves travel in a medium with velocity 2×10^8 m/s. The relative permeability of the medium is 2.0. Find the relative permittivity of the medium.

Solution: The velocity of electromagnetic wave is given by

$$v = \frac{1}{\sqrt{\mu\varepsilon}} = \frac{1}{\sqrt{\mu_0\mu_r} \; \sqrt{\varepsilon_0\varepsilon_r}}$$

$$= \frac{1}{\sqrt{\mu_0\varepsilon_0} \; \sqrt{\mu_r\varepsilon_r}}$$

But in free space

$$\frac{1}{\sqrt{\mu_0\varepsilon_0}} = c \text{ i.e. speed of light}$$

so

$$v = \frac{c}{\sqrt{\mu_r\varepsilon_r}}$$

or

$$\varepsilon_r = \frac{c^2}{v^2\,\mu_r} = \frac{\left(3\times10^8\right)^2}{\left(2\times10^8\right)^2 \times 2.0}$$

$$= 1.125$$

Example 6: The electric field of a plane electromagnetic wave in vacuum is represented by $E_x = 0$, $E_y = 0.5 \cos\left[2\pi\times10^8\left(t - \dfrac{x}{c}\right)\right]$ and $E_z = 0$.

find (i) What is the direction of wave propagation?

(ii)   What is the associated wavelength?

(iii)   Calculate the components of magnetic field associated with the wave.

[DSSCE 92 comptt)

Solution

(i)   The direction of wave propagation is +ve direction.

(ii)   Comparing the given equation

$$y = 0.5 \cos\left[2\pi \times 10^8\left(t - \frac{x}{c}\right)\right]$$

with the standard wave equation.

$$y = a \cos\left[\frac{2\pi C}{\lambda}\left(t - \frac{x}{c}\right)\right]$$

we get $$\frac{2\pi C}{\lambda} = 2\pi \times 10^8$$

or $\lambda = 3$ m

(iii)   here $B_x = 0$, $B_y = 0$ but $B_z \neq 0$

        Using $E = CB$

$$B_z = \frac{E_y}{C} = \frac{0.5\cos\left[2\pi \times 10^8\left(t - \frac{x}{c}\right)\right]}{3 \times 10^8}$$

$$= 0.17 \times 10^{-8} \cos\left[2\pi \times 10^8\left(t - \frac{x}{c}\right)\right] \text{ Tesla}$$

Example 7 : An observer is at a distance of 1.0 m from a point source of light of Power output 10^3W. Calculate the rms values of the electric and magnetic fields due to the source at the position of the observer. Assume that the source is monochromatic and radiates uniformly in all directions.   **[CEE 95]**

Solution: The intensity I of the wave at a distance r from the source is the rate of which energy from point source of power P is transferred across unit area is

$$I = \frac{P}{4\pi r^2} \qquad \qquad ...(1)$$

for any electromagnetic wave

$$I = \frac{1}{c\mu_0}E_{rms}^2$$

$$\Rightarrow \qquad E_{rms} = \sqrt{\frac{(PC\mu_0)}{4\pi r^2}}$$

$$= \sqrt{\frac{10^3 \times 3 \times 10^8 \times 4\pi \times 10^{-7}}{4\pi \times (1.0)^2}}$$

$$= 173 \ \text{Vm}^{-1}$$

so the rms value of the magnetic field

$$B_{rms} = \frac{E_{rms}}{c} = \frac{173}{3 \times 10^8}$$

$$= 5.77 \times 10^{-7} \ \text{T}.$$

Example 8 : Calculate the skin depth in aluminium at 1.8 MHz where $\sigma = 40 \times 10^6$ mho m^{-1} and $\mu = 4\pi \times 10^{-7} \text{H m}^{-1}$.

Solution: We know that

Skin depth
$$\delta = \left(\frac{2}{\mu \omega \sigma}\right)^{1/2}$$

$$= \left[\frac{2}{\mu(2\pi \nu)\sigma}\right]^{1/2} = \left(\frac{1}{\pi \nu \mu \sigma}\right)^{1/2}$$

$$= \frac{1}{\sqrt{\pi \times 1.8 \times 10^6 \times 4\pi \times 10^{-7} \times 40 \times 10^6}}$$

SUMMARY

* During the process of capacitor charging, a conduction current i exists through connecting wires and a displacement current i_d in between capacitor plates

here
$$i_d = \varepsilon_0 \frac{d\phi_e}{dt}$$

or
$$i_{df} = A\varepsilon_0 \frac{\partial E}{\partial t}$$

and displacement current density

$$\vec{J}_d = \frac{\partial \vec{D}}{\partial t}$$

- Solution of maxwell equation in free space

$$\nabla^2 \vec{E} = \mu_0 \varepsilon_0 \frac{\partial^2 \vec{E}}{\partial t^2}$$

$$\nabla^2 \vec{H} = \mu_0 \varepsilon_0 \frac{\partial^2 \vec{H}}{\partial t^2}$$

speed of light in free space

$$c = \frac{1}{\sqrt{\mu_0 \varepsilon_0}} \simeq 3\times10^8 \, m/s$$

- Solution of maxwell equation in charge free conductivity medium

$$\nabla^2 \vec{E} - \varepsilon\mu \frac{\partial^2 \vec{E}}{\partial t} - \mu\sigma \frac{\partial \vec{E}}{\partial t} = 0$$

$$\nabla^2 \vec{H} - \sigma\mu \frac{\partial^2 \vec{H}}{\partial t} - \varepsilon\mu \frac{\partial^2 \vec{H}}{\partial t^2} = 0$$

- Skin depth is defined as the distance or depth of penetration, where the amplitude of electric field reduce to $\frac{1}{\rho}$ time of the amplitude of electric field at the surface.

$$\delta =- \sqrt{\frac{2}{\mu\sigma\omega}}$$

- Electromagnetic energy density

$$U = \frac{1}{2}\left(\varepsilon_0 E^2 + \frac{B^2}{\mu_0} \right)$$

- The rate of energy transfer through per unit area is called poynting vector

$$\vec{P} = \vec{E} \times \vec{H}$$

- According to poynting theorem

Total power leasing the volume = ratio of decrease of stored electromagnetic energy – Ohmic power dissipated due to motion of charge.

$$\int_s \vec{P} \cdot \vec{ds} = -\int_V \frac{\partial U_{em}}{\partial t} dV - \int_V \left(\vec{E} \cdot \vec{J}\right) dV$$

EXERCISE

1. What do you understand by displacement current? Explain the continuity of sum of conduction and displacement currents through individually these are discontinuous.

2. State and explain Maxwell's modification of Ampere's law.

3. Use Maxwell's equation to obtain expression for the velocity of electromagnetic waves and hence show that light is an electromagnetic wave.

4. Show that in electromagnetic wave, the average electric energy density equals the average magnetic energy density.

5. State the significance of Gauss's law and maxwell's modifications of amperes law.

6. Give the maxwell's equations for good conductors and dielectric medium.

7. Derive maxwell's equation when it is passing through any free space.

8. Prove that the possible solutions for Maxwell's equations are

$$\vec{E} = \vec{E}_0 e^{i\left(\vec{k} \cdot \vec{r} - \omega t\right)}$$

and

$$\vec{H} = \vec{H}_0 e^{i\left(\vec{k} \cdot \vec{r} - \omega t\right)}$$

9. Write Maxwell's equations for charge free non-conducting medium.

10. What do you understand by skin depth. Prove that skin depth decreases with increase in frequency of electromagnetic wave and the conductivity of the medium.

11. Explain the physical significance of skin depth.

12. Prove that electromagnetic wave attenuates when it propagates through a conducting medium.

13. What is poynting vector? Given its importance.

14. Show that average value of poynting vector for a plane electromagnetic wave is $\dfrac{1}{2}\sqrt{\dfrac{\mu}{\varepsilon}}\,\mu_0^2$.

15. The voltage between the parallel plates of a 1.5 mF capacitor changes at a rate of 5 Vs^{-1} find the displacement current.

16. A 200 pf capacitor is connected to 220 V – 50 Hz a.c. mains. Fin the rms value of displacement current.

17. Parallel plate capacitor has circular plates, each of radius 5.0 cm with electric field change in the gap at the rate of 10^{12} Vm^{-1} s^{-1} Find the value of displacement current.

18. A parallel plate capacitor is made of two rectangular plates of size 35 × 20 cm^2 and separated by a distance 3.0 mm. Find the rate of change of potential difference across the plates and displacement current when it is charged by an external source at a constant charging current of 0.2 A.

19. A parallel plate air capacitor made of circular plates is being charged by constant current. If the magnetic field at a distance 20 cm from the axis of the plate in 3× 10^{-7}j, find the charging current.

20. A 200 pf parallel plate capacitor is made of circular plates each of radius 15 cm. It is connected to a (200 V – 200 Hz) a.c. supply. Find the rms value of conduction current, peak value of displacement current and the amplitude of $\vec{B}$ at a point 3.0 cm from the axis between the plates.

21. Electromagnetic waves enter a medium at relative permeability 6 and relative permittivity of 85, Find the speed of e.m. waves in the medium.

22. A plane electromagnetic wave has a maximum electric field of 4.9 × 10^{-4} V/m. What is the maximum magnetic field?

23. In a plane electromagnetic wave, the amplitude of the magnetic field is 4.9 × 10^{-6} T. Find the amplitude of the electric field and the total average energy density of the waves.

24. In a plane electromagnetic wave the electric field varies with time having an amplitude of 2.0 V/m. Find the average energy density of magnetic field and that of electric field.

25. The electric field associated with a plane electromagnetic wave is given by $E_x = 0$, $E_y = 0$, $E_z = 2.0 \cos\left[2\pi \times 10^5 \left(t - \dfrac{x}{c}\right)\right]$ with $c = 3\times10^8$ m/s. Write expressions for the components of the magnetic field of the wave.

26. Determine the propagation constant k for a material having $\mu_r = 1.2$, $\varepsilon_r = 6$ and $\sigma = 0.35$ p s/m, if the wave frequency is 1.8 MHz.

27. Calculate the skin depth for copper at a frequency of 40 MHz if its $\mu = 4\pi \times 10^{-7}$ m / m and $\sigma = 6.8 \times 10^7$ mho/m.

28. A 200 MHz plane wave penetrates through a medium of conductivity $\sigma = 10^5$ mho or s/m, $\varepsilon_2 = \mu_r = 1$. Calculate skin depth and also depth at which the wave amplitude decreases to 13.5% of its initial value.

Module – 5

Quantum Mechanics

Generalized coordinates, Lagrange's Equations of motion and Lagrangian, Generalized Force, Potential, Momenta and Energy, Hamilton's equation of motion and Hamiltonian, Properties of Hamiltonian and Hamilton's equation of motion (course should be discussed along with physical problems of 1-D motion).

Concept of Probability and probability density, operators, Commutator, Formulation of quantum mechanics and basic postulates, operator correspondence, time dependent Schrödinger equation, Formulation of time independent Schrödinger equation by method of separation of variables, physical interpretation of wave function Ψ(normalization and probability interpretation), expectation values, Application of Schrödinger equation – Particle sin an infinite square well potential (1-D and 3-D potential well) discussion on degenerate levels.

INTRODUCTION

The branch of physics which deals with physical objects in motion and at rest under the influence of external and internal interaction is called mechanics. Newtonian mechanics deals with Newton's laws and its consequences whereas analytical mechanics deals with related schemes given by D'Alembert, Lagrange, and Hamilton etc. Classical mechanics was remarkably successful to deal with the motion of massive particles moving with slow speed. Many laws of classical mechanics, like conservation laws of energy, linear momentum and angular momentum are of universal importance. In Newtonian mechanics all the laws are valid only in inertial frames which move with constant velocities or with zero velocity i.e. at rest. Here to write the differential equations of motion a particular coordinate system is to be selected. Here the equation of motion changes its form according to the selected coordinate system hence Newtonian mechanics is not invariant under any coordinate system. In Newtonian mechanics, the solution of a equation to know the nature of motion is very difficult to find because Newton's laws require complete specifications of all the forces at all instants of time acting on a body. Further if there is any restriction on the movement of a body, then Newtonian mechanics cannot deal with that.

Constraint (i.e. restriction)

The motion of a particle or system of particles is controlled by one or more conditions. These limitations on the motion of a system are called constraints and such motion is called constrained motion.

Types of constraints

Depending on the nature of condition, constraints are classified as under-

(i)   Holonomic or Integrable constraints

1

When the constraints relation is independent of velocity and can be expressed in the form of an algebraic equation $f(\vec{r_1}, \vec{r_2}........\vec{r_n}, t) = 0$ then it is called holonomic or Integrable or geometric constraint and when constraint cannot be expressed in such a manner, it is called non-holonomic constraints, which depends on velocity.

(ii)   Scleronomic (or stationary) and Rheonomic Constraints

When constraint relation does not depend on time it is Scleronomic otherwise it is called Rheonomic constraint.

(iii)   Bilateral and unilateral constraints

When at the point of constraint, forward and backward motion is possible, then constraint is called bilateral otherwise it called unilateral.

(iv)   Conservative and dissipative constraints

If the total mechanical energy of the system remains conserved during constrained motion such that work done by constrained force are zero, then it is called conservative constraint otherwise it is called dissipative constraint.

Degrees of Freedom

These are the minimum number of independent variables or coordinates required to specify the position of a dynamic system without violating any constraint applied on that. The configuration of N free particles has 3N number of degree of freedom, as per 3N dimensional space which is called configuration space of the system.

The number of coordinates, required to specify a dynamical system, becomes less, when any constraint works on the system. Hence we can define degree of freedom of dynamical system as the minimum number of independent coordinates or parameters required to specify the system with constraints. If N free particles has m number of constraints then the number of independent coordinates will be n=3N-m.

For example:

- For a particle in space, degree of freedom is 3 as three coordinates (x,y,z) are required to specify its position.
- For any particle in a plane, it has two degree of freedom as we require (x,y) coordinates to specify its position.
- For any particle moving on the spherical system, degree of freedom are 3x1-1 =2 as the constraint =1.

Generalized coordinates

We know that mechanical problems are solved much easily when they are expressed in Cartesian coordinates but in many specific conditions use of Cartesian coordinates does not help. For example for particle which is under the effect of a fixed attracting centre like motion of electron around nucleus etc., it is hard to specify the problem but it will be easy if problem statements uses spherical coordinate system

2

(r,θ,ϕ). Further configuration of any system is restricted by constraints, making coordinates restricted according to them and hence total number of degree of freedom as $3N\text{-}m$ for system of N particles with m number of constraints working on the system.

Hence to specify the configuration of a dynamical system ($3N\text{-}m$) number of independent coordinates are to be introduced. Hence the name generalized coordinates is provided to a set of independent coordinates sufficient to describe completely the state of configuration of a dynamical system, represented by $q_1, q_2 \ldots \ldots q_k \ldots \ldots q_n$ where n is the total number of generalized coordinates. Actually, it is the minimum number of coordinates needed to describe the motion of the system. For example any particle constrained to move on the circumference of a circle only one coordinate $q_1 = \theta$ is sufficient to explain its motion. It is not compulsory that these are rectangular, spherical or cylindrical. The quantities like length, length^2, angle etc may be used as generalized coordinates provided these should completely describe the system.

For a system of N particles, if x_i, y_i, z_i are the Cartesian coordinates of the ith particle, then

$$x_i = x_i(q_1, q_2 \ldots \ldots q_k \ldots q_n, t)$$
$$y_i = y_i(q_1, q_2 \ldots \ldots q_k \ldots q_n, t)$$
$$z_i = z_i(q_1, q_2 \ldots \ldots q_k \ldots q_n, t)$$
$$\vec{r_i} = \vec{r_i}(q_1, q_2 \ldots \ldots q_k \ldots q_n, t)$$

Principle of Virtual Work

To study the system under motion, we can take small instantaneous change in the position vectors of the particles of the system e.g. virtual displacement, denoted by $\delta \vec{r_i}$, representing only change in position coordinates, not time.

$$\delta \vec{r_i} = \vec{r_i}(q_1, q_2 \ldots \ldots q_n) \qquad\qquad \ldots\ldots\ldots\ldots\ldots(i)$$

Under equilibrium, total force is zero i.e. $\vec{F_i} = 0$ (i=1,2…..N) and the virtual work will be zero

i.e. $\delta W_i = \vec{F_i}.\delta \vec{r_i} = 0$

And $\delta W_i = \sum\limits_{i=1}^{N} \vec{F_i}.\delta \vec{r_i} = 0$ $\ldots\ldots\ldots\ldots\ldots(ii)$

Equation (ii) represent the principle of virtual work, which states that in the position of equilibrium the work done in arbitrary virtual displacement is zero.

Here the total force is given by $\vec{F_i} = \vec{F_i^a} + \vec{F_i^c}$ where $\vec{F_i^a}$ is the applied force and $\vec{F_i^c}$ is the force due to constraints. So from (ii)

$$\delta W_i = \sum_{i=1}^{N} (\overrightarrow{F_i^a} + \overrightarrow{F_i^c}).\delta \overrightarrow{r_i} = 0$$

$$or \sum_{i=1}^{N} \overrightarrow{F_i^a}.\delta \overrightarrow{r_i} + \sum_{i=1}^{N} \overrightarrow{F_i^c}.\delta \overrightarrow{r_i} = 0$$

In case where the work done due to forces of constraints are zero, then

$$\boxed{\delta W_i = \sum_{i=1}^{N} \overrightarrow{F_i^a}.\delta \overrightarrow{r_i} = 0}$$ ....(iii)

Hence in equilibrium the virtual work of applied force is zero.

D'Alembert's Principle

According to Newton's second law

$$\overrightarrow{F_i} = \frac{\partial \overrightarrow{p_i}}{\partial t} = \dot{p_i}$$ or $$\overrightarrow{F_i} - \dot{p_i} = 0$$ (i=1,2.....N)

And for virtual displacement $\delta \overrightarrow{r_i}$

$$\delta W_i = \sum_{i=1}^{N} (\overrightarrow{F_i} - \dot{p_i}).\delta \overrightarrow{r_i} = 0$$

But $$\overrightarrow{F_i} = \overrightarrow{F_i^a} + \overrightarrow{F_i^c}$$ so

$$\sum_{i=1}^{N} (\overrightarrow{F_i^a} + \overrightarrow{F_i^c} - \dot{p}).\delta \overrightarrow{r_i} = 0$$ or $$\sum_{i=1}^{N} (\overrightarrow{F_i^a} - \dot{p}).\delta \overrightarrow{r_i} + \sum_{i=1}^{N} \overrightarrow{F_i^c}.\delta \overrightarrow{r_i} = 0$$

Or assuming that work done of constraints is zero so

$$\sum_{i=1}^{N} (\overrightarrow{F_i^a} - \dot{p}).\delta \overrightarrow{r_i} = 0$$ ....(iv)

Eq (iv) represents the D'Alembert's principle, according to which the sum of work done due to applied forces and due to the forces of inertia for any virtual displacement of the system for each instant of time, is equal to zero, for the actual motion of any system of particles under constraint motion.

Advantages of generalized coordinates

4

Generalized coordinates eliminates the dependence of coordinate system for describing the configuration of any dynamical system and these remain independent to each other making possibility of calculating the individual variations.

Some generalized physical quantities

(i)   Generalized velocity

It is the time derivative of generalized coordinate. If q_j is the generalized coordinate of a system, then generalized velocity is

$$\dot{q}_j = \frac{\partial q_j}{\partial t} \qquad (j=1,2\ldots\ldots n) \qquad\qquad \ldots\ldots\ldots\ldots\ldots(i)$$

For a constrained system of N particles, the position vector $\vec{r}_i$ is

$$\vec{r}_i = \vec{r}_i(q_1, q_2 \ldots\ldots\ldots q_n, t)$$

So $d\vec{r}_i = \sum_{i=1}^{n} \dfrac{\partial \vec{r}_i}{\partial q_j} dq_j + \dfrac{\partial \vec{r}_i}{\partial t} dt$

And $\boxed{\vec{v}_i = \dot{\vec{r}}_i = \sum_{i=1}^{n} \dfrac{\partial \vec{r}_i}{\partial q_j} \dot{q}_j + \dfrac{\partial \vec{r}_i}{\partial t}}$ where $\dot{q}_j = \dfrac{dq_j}{dt}$ $\ldots\ldots\ldots\ldots\ldots(ii)$

(ii)   Generalized force

We can write position vector $\vec{r}_i$ of a holonomic system of N particles with m constraints can be written as

$$\vec{r}_i = \vec{r}_i(q_1, q_2 \ldots\ldots\ldots q_n, t) = \vec{r}_i\left(q_j, t\right)$$

Or $\delta \vec{r}_i = \dfrac{\partial \vec{r}_i}{\partial q_1}\delta q_1 + \dfrac{\partial \vec{r}_i}{\partial q_2}\delta q_2 + \ldots\ldots \dfrac{\partial \vec{r}_i}{\partial q_t}\delta q_t$

But for virtual displacement $\delta \vec{r}_i = d\vec{r}_i\big|_{dt=0}$

$$\delta \vec{r}_i = \sum_{j=1}^{n} \frac{\partial \vec{r}_i}{\partial q_j}\delta q_j \qquad \text{as } \delta t = 0$$

Here δq_j is generalized displacement and the total virtual work done in terms of generalized coordinates, due to applied force is

$$\sum_{i=1}^{N} \vec{F}_i^a . \delta \vec{r}_i = \sum_{j=1}^{n}\left(\sum_{i=1}^{N} \vec{F}_i^a . \frac{\partial \vec{r}_i}{\partial q_j}\right)\delta q_j = \sum_{j=1}^{n} Q_j \delta q_j$$

Where Q_j is the jth component of generalized force and because generalized coordinates need not have the dimensions of length, so generalized force Q_j does

5

not necessarily have the dimensions of force but $\sum_{j=1}^{n} Q_j \delta q_j$ must have dimensions of work.

(iii)   Generalized Force in Conservative system

A conservative system is that where its total energy is conserved. The conservative force $\overrightarrow{F_i}$ can be expressed in terms of potential $V(q_j)$ as

$$\overrightarrow{F_i} = -\overrightarrow{\nabla}_i V \qquad \qquad \qquad ....(i)$$

And generalized force $Q_j = \sum_{i=1}^{N} \overrightarrow{F_i^a} . \frac{\partial \overrightarrow{r_i}}{\partial q_j} = -\sum_{i=1}^{N} \overrightarrow{\nabla}_i V . \frac{\partial \overrightarrow{r_i}}{\partial q_j}$

Or
$$Q_j = -\sum_{i=1}^{N} \frac{\partial V}{\partial \overrightarrow{r_i}} . \frac{\partial \overrightarrow{r_i}}{\partial q_j} \qquad \qquad ....(ii)$$

Now for any conservative system, scalar potential V is a function of position only, so $\boxed{Q_j = -\dfrac{\partial V}{\partial q_j}}$ $\qquad \qquad ....(iii)$

Eq.(iii) represents the generalized force for any conservative system.

(iv)   Generalized potential

Generalized potential is velocity dependent and produces generalized force. For any conservative system $\overrightarrow{F_i} = -\overrightarrow{\nabla}_i V$ Where V is scalar potential which depends on r.

So $V = V(r_1, r_2, ...r_n) = V(q_1, q_2, ...q_n)$ $\qquad \qquad ....(i)$

And $dV = \sum_{i=1}^{N} \left[\frac{\partial V}{\partial x_i} dx_i + \frac{\partial V}{\partial y_i} dy_i + \frac{\partial V}{\partial z_i} dz_i \right]$

$$= \sum_{i=1}^{N} \overrightarrow{\nabla}_i V . d\overrightarrow{r_i} \qquad \qquad ....(ii)$$

So $\dfrac{dV}{dq_j} = \sum_{i=1}^{N} \overrightarrow{\nabla}_i V . \dfrac{\partial \overrightarrow{r_i}}{\partial q_j}$ $\qquad \qquad ....(iii)$

But when system is not conservative, potential depends on the generalized velocity $\dot{q}_j$

So $Q_j = -\dfrac{\partial V}{\partial q_j} + \dfrac{d}{dt}\left(\dfrac{\partial U}{\partial \dot{q}_j} \right)$ $\qquad \qquad ....(iv)$

6

Here $U = U(q_j, \dot{q}_j)$ is called velocity dependent potential or generalized potential.

(v)   Generalized Kinetic Energy

We know that kinetic energy of ith particle of mass m_i is $\frac{1}{2}m_i\dot{r}_i^2$ and for any system of N particles

$$T = \sum_{i=1}^{N}\frac{1}{2}m_i\dot{r}_i^2 = \sum_{i=1}^{N}\frac{1}{2}m_i\vec{\dot{r}}_i.\vec{\dot{r}}_i \qquad\qquad ....(i)$$

Where $\vec{r}_i = \vec{r}_i(q_1, q_2.............q_n, t) = r_i(q_j, t)$

So $\vec{\dot{r}}_i = \sum_{j=1}^{n}\frac{\partial \vec{r}_i}{\partial q_j}\dot{q}_j + \frac{\partial \vec{r}_i}{\partial t} \qquad\qquad ....(ii)$

Hence from (i)

$$T = \sum_{i=1}^{N}\frac{1}{2}m_i\left[\sum_{j=1}^{n}\frac{\partial \vec{r}_i}{\partial q_j}\dot{q}_j + \frac{\partial \vec{r}_i}{\partial t}\right]\left[\sum_{k=1}^{n}\frac{\partial \vec{r}_i}{\partial q_k}\dot{q}_k + \frac{\partial \vec{r}_i}{\partial t}\right]$$

$$= \sum_{i=1}^{N}\sum_{j=1}^{n}\sum_{k=1}^{n}\left[\frac{1}{2}m_i\frac{\partial \vec{r}_i}{\partial q_j}.\frac{\partial \vec{r}_i}{\partial q_k}\dot{q}_j\dot{q}_k + \frac{1}{2}m_i\frac{\partial \vec{r}_i}{\partial q_k}.\frac{\partial \vec{r}_i}{\partial t}\dot{q}_k + \frac{1}{2}m_i\frac{\partial \vec{r}_i}{\partial t}.\frac{\partial \vec{r}_i}{\partial q_j}\dot{q}_j + \frac{1}{2}m_i\frac{\partial \vec{r}_i}{\partial t}.\frac{\partial \vec{r}_i}{\partial t}\right]$$

$$= \sum_{i=1}^{N}\sum_{j=1}^{n}\sum_{k=1}^{n}\frac{1}{2}m_i\frac{\partial \vec{r}_i}{\partial q_j}.\frac{\partial \vec{r}_i}{\partial q_k}\dot{q}_j\dot{q}_k + \sum_{i,j}\frac{1}{2}(2m_i)\frac{\partial \vec{r}_i}{\partial t}.\frac{\partial \vec{r}_i}{\partial q_j}\dot{q}_j + \sum_{i=1}^{N}\frac{1}{2}m_i\left(\frac{\partial \vec{r}_i}{\partial t}\right)^2$$

Now consider $A_{j,k} = \sum_{i=1}^{N}\frac{1}{2}m_i\frac{\partial \vec{r}_i}{\partial q_j}.\frac{\partial \vec{r}_i}{\partial q_k}$, $A_j = \sum_{i=1}^{N}m_i\frac{\partial \vec{r}_i}{\partial q_j}.\frac{\partial \vec{r}_i}{\partial t}$.

And $A = \sum_{i=1}^{N}\frac{1}{2}m_i\left(\frac{\partial \vec{r}_i}{\partial t}\right)^2$

Hence using these in above for T

$$\boxed{T = \sum_{j=1}^{n}\sum_{k=1}^{n}A_{j,k}\dot{q}_j\dot{q}_k + \sum_{j=1}^{n}A_j\dot{q}_j + A} \qquad\qquad ....(iii)$$

Eq (iii) represents the kinetic energy which is a quadratic function of generalized velocities.

$r_i = r_i(q_j)$ then transformation equations are independent of time i.e. $\dfrac{\partial \vec{r_i}}{\partial t} = 0$

and $A_j = A = 0$ then kinetic energy equation reduces to

$$T = \sum_{j=1}^{n} \sum_{k=1}^{n} A_{j,k} \dot{q}_j \dot{q}_k$$ ....(iv)

(vi)   Generalized Momentum

Generalized momentum is defined as the partial derivative of total energy (E) of the system with respect to generalized velocities.

$p_k = (p_1, p_2............p_n)$ where n is the number of generalized coordinates i.e. degree of freedom.

$p_j = \dfrac{\partial E}{\partial \dot{q}_j}$ where $j = 1,2..............n$

$= \dfrac{\partial (T+V)}{\partial \dot{q}_j}$ where t is kinetic energy and V is the potential energy of the system

Here V depends only on q_j and so $\dfrac{\partial V}{\partial \dot{q}_j} = 0$

Hence $p_j = \dfrac{\partial T}{\partial \dot{q}_j}$ ....(v)

Eq (v) represents the expression for generalized momentum.

Lagrangian Coordinates

In Newtonian Mechanics, restriction towards the coordinate system was observed due to its confinement about a particular coordinate system, but in many cases this is not required as this does not help to find solutions to certain problems specified in other coordinate systems. Lagrangian used generalized coordinates, which are independent of any particular coordinate system, to overcome these difficulties of Newtonian mechanics.

Lagrange's equations from D'Alembert's Principle

Let us assume a system of N particles for which the transformation equations for the position vectors can be written as

$$\vec{r_i} = \vec{r_i}\left(q_1, q_2............q_k, t\right)$$ ....(i)

Here q_k is the generalized coordinate with $k = 1,2,3, ...n$ and t is time.

Differentiating (i) w.r.t. t

$$\vec{v}_i = \dot{\vec{r}}_i = \sum_{k=1}^{n} \frac{\partial \vec{r}_i}{\partial q_k} \dot{q}_k + \frac{\partial \vec{r}_i}{\partial t} \qquad\qquad \text{..............(ii)}$$

Here $\dot{q}_k$ are the generalized velocities.

Now the virtual displacement is given by

$$\delta \vec{r}_i = \frac{\partial \vec{r}_i}{\partial q_1} \delta q_1 + \frac{\partial \vec{r}_i}{\partial q_2} \delta q_2 + \text{............} + \frac{\partial \vec{r}_i}{\partial q_k} \delta q_k + \text{.......} + \frac{\partial \vec{r}_i}{\partial q_n} \delta q_n$$

$$\delta \vec{r}_i = \sum_{k=1}^{n} \frac{\partial \vec{r}_i}{\partial q_k} \delta q_k \qquad\qquad \text{..............(iii)}$$

Since virtual displacement does not depend on time.

According to D'Alembert's principle, virtual work is given by

$$\sum_{i=1}^{N} \left(\vec{F}_i - \dot{\vec{p}}_i \right) . \delta \vec{r}_i = 0 \qquad\qquad \text{..............(iv)}$$

With $\displaystyle \sum_{i=1}^{N} \vec{F}_i . \delta \vec{r}_i = \sum_{i=1}^{N} \vec{F}_i . \sum_{k=1}^{n} \frac{\partial \vec{r}_i}{\partial q_k} \delta q_k = \sum_{k=1}^{n} \sum_{i=1}^{N} \left[\vec{F}_i . \frac{\partial \vec{r}_i}{\partial q_k} \right] \delta q_k$

$$= \sum_{k=1}^{n} G_k \delta q_k \qquad\qquad \text{..............(v)}$$

where $\displaystyle G_k = \sum_{i=1}^{N} \left[\vec{F}_i . \frac{\partial \vec{r}_i}{\partial q_k} \right] = \sum_{i=1}^{N} \left[F_{xi} . \frac{\partial x_i}{\partial q_k} + F_{yi} . \frac{\partial y_i}{\partial q_k} + F_{zi} . \frac{\partial z_i}{\partial q_k} \right]$ ....(vi)

are called the generalized force. This is associated with generalize coordinates q_k.

and $\displaystyle \sum_{i=1}^{N} \dot{\vec{p}}_i . \delta \vec{r}_i = \sum_{i=1}^{N} m_i \ddot{\vec{r}}_i . \sum_{k=1}^{n} \frac{\partial \vec{r}_i}{\partial q_k} \delta q_k = \sum_{k=1}^{n} \left[\sum_{i=1}^{N} m_i \ddot{\vec{r}}_i . \frac{\partial \vec{r}_i}{\partial q_k} \right] \delta q_k$ ....(vii)

but $\displaystyle \sum_{i=1}^{N} m_i \ddot{\vec{r}}_i . \frac{\partial \vec{r}_i}{\partial q_k} = \sum_{i=1}^{N} \left[\frac{d}{dt} \left(m_i \dot{\vec{r}}_i . \frac{\partial \vec{r}_i}{\partial q_k} \right) - m_i \dot{\vec{r}}_i . \frac{d}{dt} \left(\frac{\partial \vec{r}_i}{\partial q_k} \right) \right]$ ....(viii)

and it is easy to prove that $\displaystyle \frac{d}{dt} \left(\frac{\partial \vec{r}_i}{\partial q_k} \right) = \frac{\partial}{\partial q_k} \left(\frac{d \vec{r}_i}{dt} \right) = \frac{\partial \vec{v}_i}{\partial q_k}$

and $\dfrac{\partial \vec{r_i}}{\partial q_k} = \dfrac{\partial \vec{v_i}}{\partial \dot{q}_k}$

hence $\displaystyle\sum_{i=1}^{N} m_i \vec{\ddot{r_i}} \cdot \dfrac{\partial \vec{r_i}}{\partial q_k} = \sum_{i=1}^{N}\left[\dfrac{d}{dt}\left(m_i\vec{v_i}\cdot\dfrac{\partial \vec{v_i}}{\partial \dot{q}_k}\right) - m_i\vec{v_i}\cdot\left(\dfrac{\partial \vec{v_i}}{\partial q_k}\right)\right]$ ....(ix)

using (ix) in (vii)

$$\sum_{i=1}^{N}\vec{p_i}.\delta\vec{r_i} = \sum_{k=1}^{n}\sum_{i=1}^{N}\left[\dfrac{d}{dt}\left(m_i\vec{v_i}\cdot\dfrac{\partial \vec{v_i}}{\partial \dot{q}_k}\right) - m_i\vec{v_i}\cdot\left(\dfrac{\partial \vec{v_i}}{\partial q_k}\right)\right]\delta q_k$$

$$= \sum_{k=1}^{n}\left[\dfrac{d}{dt}\left\{\dfrac{\partial}{\partial \dot{q}_k}\left(\sum_{i=1}^{N}\dfrac{1}{2}m_i\left(\vec{v_i}.\vec{v_i}\right)\right)\right\} - \dfrac{\partial}{\partial q_k}\left\{\sum_{i=1}^{N}\dfrac{1}{2}m_i\left(\vec{v_i}.\vec{v_i}\right)\right\}\right]\delta q_k$$

$$= \sum_{k=1}^{n}\left[\dfrac{d}{dt}\left(\dfrac{\partial T}{\partial \dot{q}_k}\right) - \dfrac{\partial T}{\partial q_k}\right]\delta q_k$$ ....(x)

Where $T = \displaystyle\sum_{i=1}^{N}\dfrac{1}{2}m_i\left(\vec{v_i}.\vec{v_i}\right) = \sum_{i=1}^{N}\dfrac{1}{2}m_i v_i^2$ is the kinetic energy of the the system.

Hence eq (v) becomes

$$\sum_{k=1}^{n}\left[\left\{\dfrac{d}{dt}\left(\dfrac{\partial T}{\partial \dot{q}_k}\right) - \dfrac{\partial T}{\partial q_k}\right\} - G_k\right]\delta q_k = 0$$ ....(xi)

Virtual displacement is independent of δq_k because constraints are holonomic.

Hence $\left\{\dfrac{d}{dt}\left(\dfrac{\partial T}{\partial \dot{q}_k}\right) - \dfrac{\partial T}{\partial q_k}\right\} - G_k = 0$ or $\boxed{\dfrac{d}{dt}\left(\dfrac{\partial T}{\partial \dot{q}_k}\right) - \dfrac{\partial T}{\partial q_k} = G_k}$ ....(xii)

Eq (xii) represents the general form of Lagrange's equation.

For a conservative system

$$G_k = -\dfrac{\partial V}{\partial q_k}$$ ....(xiii)

So from eq (xii)

$$\dfrac{d}{dt}\left(\dfrac{\partial T}{\partial \dot{q}_k}\right) - \dfrac{\partial T}{\partial q_k} = -\dfrac{\partial V}{\partial q_k}$$ or $\boxed{\dfrac{d}{dt}\left(\dfrac{\partial T}{\partial \dot{q}_k}\right) - \dfrac{\partial(T-V)}{\partial q_k} = 0}$ ....(xiv)

As scalar potential V is the function of generalized coordinate q_k only and does not depend upon generalized velocities so eq (xiv) can be written as

$$\frac{d}{dt}\left(\frac{\partial(T-V)}{\partial \dot{q}_k}\right) - \frac{\partial(T-V)}{\partial q_k} = 0$$

Assuming a new function L=T – V

So $\boxed{\dfrac{d}{dt}\left(\dfrac{\partial L}{\partial \dot{q}_k}\right) - \dfrac{\partial L}{\partial q_k} = 0}$ for k =1,2,…………..n   ……………..(xv)

Eq (xv) represents Lagrange's equation for conservative system. There is one equation for every generalized coordinate, and there are n in numbers.

Application of Lagrange's Equation

1. Newton's equation of motion

    Here q_1=x, q_2=y and q_3=z   and G_1=F_x, G_2=F_y and G_3=F_z
    As general form of Lagrange's equation is

    $$\frac{d}{dt}\left(\frac{\partial T}{\partial \dot{q}_k}\right) - \frac{\partial T}{\partial q_k} = G_k$$ ……………..(i)

    And kinetic energy $T = \frac{1}{2}m\left[\dot{x}^2 + \dot{y}^2 + \dot{z}^2\right]$

    For x coordinate only, using eq (i)

    $$\frac{d}{dt}\left(\frac{\partial T}{\partial \dot{x}}\right) - \frac{\partial T}{\partial x} = F_x$$ ……………..(ii)

    But $\dfrac{\partial T}{\partial x} = 0$ and $\dfrac{\partial T}{\partial \dot{x}} = m\dot{x}$

    So from (ii) $\dfrac{d}{dt}(m\dot{x}) = F_x$

    Or $F_x = \dfrac{dp_x}{dt}$ where p_x is the x-component of momentum.

    Similarly, we can get $F_y = \dfrac{dp_y}{dt}$ and $F_z = \dfrac{dp_z}{dt}$

    So $\boxed{\vec{F} = \dfrac{d\vec{p}}{dt}}$ ……………..(iii)

    Eq (iii) represents the Newton's equation of motion.

2. Motion of free particle

A free particle is that which is not acted upon by any external force, so the potential energy V becomes zero.

Hence Lagrangian of this particle L = T – V

$$L = \frac{1}{2}m\left(\dot{x}^2 + \dot{y}^2 + \dot{z}^2\right)$$

Hence Lagrangian equation of motions

$$\frac{d}{dt}\left(\frac{\partial L}{\partial \dot{x}}\right) = \frac{\partial L}{\partial x} \quad \text{or } m\ddot{x} = 0$$

$$\frac{d}{dt}\left(\frac{\partial L}{\partial \dot{y}}\right) = \frac{\partial L}{\partial y} \quad \text{or } m\ddot{y} = 0$$

$$\frac{d}{dt}\left(\frac{\partial L}{\partial \dot{z}}\right) = \frac{\partial L}{\partial z} \quad \text{or } m\ddot{z} = 0$$

These are equation of motion for any free particle.

3. Simple Pendulum

Let us consider a simple pendulum of mass m and effective length l suspended from C. here the angle θ between the rest position and deflected position of the bob is considered as generalized coordinate.

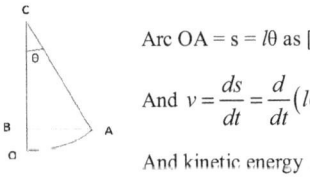

Arc OA = s = $l\theta$ as $\left[\theta = \frac{s}{l}\right]$

And $v = \frac{ds}{dt} = \frac{d}{dt}(l\theta) = \frac{l d\theta}{dt} = l\dot{\theta}$

And kinetic energy $T = \frac{1}{2}mv^2 = \frac{1}{2}ml^2\dot{\theta}^2$

And potential energy

$$V = mg(OB)$$
$$= mg(OC - BC) = mg(l - l\cos\theta) = mgl(1 - \cos\theta)$$

Hence Lagrangian   L = T – V

$$L = \frac{1}{2}ml^2\dot{\theta}^2 - mgl\left(1 - \cos\theta\right)$$

And $\frac{\partial L}{\partial \theta} = -mgl\sin\theta$ and $\frac{\partial L}{\partial \dot{\theta}} = ml^2\dot{\theta}$

Using the Lagrange's equation $\frac{d}{dt}\left(\frac{\partial L}{\partial \dot{\theta}}\right) - \frac{\partial L}{\partial \theta} = 0$

$$\frac{d}{dt}\left[ml^2\dot{\theta}\right]+mgl\sin\theta=0$$

Or $ml^2\ddot{\theta}+mgl\sin\theta=0$

Or $\boxed{\ddot{\theta}+\dfrac{g}{l}\sin\theta=0}$

Above equation represents the equation of motion of a simple pendulum. For small oscillations $\sin\theta\simeq\theta$

So $\ddot{\theta}+\dfrac{g}{l}\theta=0$

Comparing with general equation of SHM $\ddot{\theta}+\omega^2\theta=0$

$\omega=\sqrt{\dfrac{g}{l}}$ and time period of oscillation $\boxed{T=\dfrac{2\pi}{\omega}=2\pi\sqrt{\dfrac{1}{g}}}$

4. Simple Harmonic Oscillator

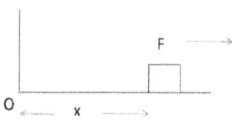

When the free vibrations of vibrating system is simple harmonic, system is called harmonic oscillator.

Here magnitude of restoring force $F\propto-x$ where x is the displacement of the body at any instant from the fixed point.

So F = -kx   where k is th spring constant or force constant of spring.

The kinetic energy of this oscillator $T=\dfrac{1}{2}m\dot{x}^2$

and potential energy $V=-\int Fdx=-\int_{\infty}^{x}-kxdx=\dfrac{1}{2}kx^2$

so Lagrangian   L $=T-V$

$$=\dfrac{1}{2}m\dot{x}^2-\dfrac{1}{2}kx^2$$

So $\dfrac{\partial L}{\partial\dot{x}}=m\dot{x}$ and $\dfrac{\partial L}{\partial x}=-kx$

Now using Lagrange's equation $\dfrac{d}{dt}\left(\dfrac{\partial L}{\partial\dot{x}}\right)-\dfrac{\partial L}{\partial x}=0$

$$\frac{d}{dt}(m\dot{x}) + kx = 0 \qquad \text{or} \qquad m\ddot{x} + kx = 0 \qquad \text{or} \qquad \ddot{x} + \frac{k}{m}x = 0$$

Comparing this with general equation of SHM $\ddot{x} + \omega^2 x = 0$

Hence time period of the harmonic oscillator

$$\boxed{T = 2\pi\sqrt{\frac{m}{k}}}$$

5. Atwood Machine

The Atwood machine is an example of a conservative system, with holonomic constraint, where two masses m_1 and m_2 are connected by an inextensible and massless string of length l through a small, massless and frictionless pulley.

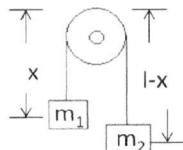

Here the only independent coordinate is x and the velocities of two masses v_1 and v_2 as $v_1 = \frac{dx}{dt} = \dot{x}$ and $v_2 = \frac{d(l-x)}{dt} = -\dot{x}$

Hence kinetic energy $T = \frac{1}{2}m_1\dot{x}^2 + \frac{1}{2}m_2\dot{x}^2 = \frac{1}{2}(m_1 + m_2)\dot{x}^2$

And potential energy of the system with reference to the pulley is $V = -m_1 gx - m_2 g(l - x)$

so the Lagrangian L = T – V $= \frac{1}{2}(m_1 + m_2)\dot{x}^2 + m_1 gx + m_2 g(l - x)$

now $\quad \frac{\partial L}{\partial \dot{x}} = (m_1 + m_2)\dot{x} \qquad$ and $\qquad \frac{\partial L}{\partial x} = (m_1 - m_2)g$

using the Lagrange's equation $\frac{d}{dt}\left(\frac{\partial L}{\partial \dot{x}}\right) - \frac{\partial L}{\partial x} = 0$

or $\qquad\qquad (m_1 + m_2)\ddot{x} - (m_1 - m_2)g = 0$

or $\qquad\qquad \boxed{\ddot{x} = \frac{(m_1 - m_2)}{m_1 + m_2}g} \qquad$ represents the equation of motion.

Here tension of the rope, the force of constraint is not used in Lagrangian formulation. If $m_1 > m_2$, mass m_1 descends and if $m_1 < m_2$ mass m_1 ascends.

6. Central Force Problem

When a particle move in such a manner that force on it is directed towards a fixed centre then that is called central force field system and the force is known as central force, which is conservative and the motion remains in a plane.

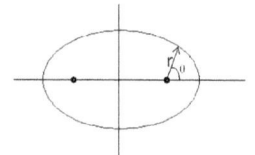

Let (r,θ) be the coordinates of the plane for particle of mass m.

Kinetic energy $T = \dfrac{1}{2}m\left(\dot{r}^2 + r^2\dot{\theta}^2\right)$

And Lagrangian $L = T - V$

$$= \frac{1}{2}m\left(\dot{r}^2 + r^2\dot{\theta}^2\right) - V(r)$$

Where V(r) is the potential energy in the central force field.

Now $\dfrac{\partial L}{\partial \dot{r}} = m\dot{r}$ and $\dfrac{\partial L}{\partial r} = mr\dot{\theta}^2 - \dfrac{\partial V}{\partial r}$ and $\dfrac{\partial L}{\partial \dot{\theta}} = mr^2\dot{\theta}$

And $\dfrac{\partial L}{\partial \theta} = 0$

Hence equations of motion are

$\dfrac{d}{dt}\left(\dfrac{\partial L}{\partial \dot{r}}\right) - \dfrac{\partial L}{\partial r} = 0$ and $\dfrac{d}{dt}\left(\dfrac{\partial L}{\partial \dot{\theta}}\right) - \dfrac{\partial L}{\partial \theta} = 0$

Or $\dfrac{d}{dt}(m\dot{r}) - mr\dot{\theta}^2 + \dfrac{\partial V}{\partial r} = 0$ and $\dfrac{d}{dt}\left(mr^2\dot{\theta}\right) = 0$

Or $m\ddot{r} - mr\dot{\theta} + \dfrac{\partial V}{\partial r} = 0$ and $\dfrac{d}{dt}\left(mr^2\dot{\theta}\right) = 0$

And for conservative force, due to attractive inverse square law $F = -\dfrac{\partial V}{\partial r} = -\dfrac{k}{r^2}$

Hence $\boxed{m\ddot{r} - mr\dot{\theta}^2 + \dfrac{k}{r^2} = 0}$ and $\dfrac{d}{dt}\left(mr^2\dot{\theta}\right) = 0$ or $\boxed{r\ddot{\theta} + 2\dot{r}\dot{\theta} = 0}$

Above equations represent the motion of particle under central force field.

7. Compound Pendulum

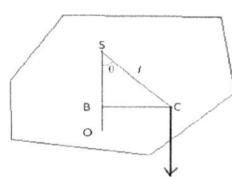

A compound pendulum is a rigid body, which is capable of oscillating in a vertical plane above a fixed horizontal axis.

Suppose the compound pendulum be suspended from S with C as centre of mass, with oscillations in the vertical plane.

Moment of Inertia of the pendulum about the axis of rotation through S is given by $I = I_c + Ml^2 = M\left(k^2 + l^2\right)$ where $I_c = Mk^2$ is the moment of Inertia about a parallel axis through C and l is the distance between centre of suspension and centre of mass.

If θ is the instantaneous angle then

kinetic energy of the oscillating system is $T = \frac{1}{2}I\dot{\theta}^2 = \frac{1}{2}M\left(k^2 + l^2\right)\dot{\theta}^2$

and potential energy $\qquad\qquad V = -Mgl\cos\theta$

and Lagrangian L = T – V $\qquad L = \frac{1}{2}M\left(k^2 + l^2\right)\dot{\theta}^2 + Mgl\cos\theta$

so $\dfrac{\partial L}{\partial \theta} = -Mgl\sin\theta$ $\qquad$ and $\qquad \dfrac{\partial L}{\partial \dot{\theta}} = M\left(k^2 + l^2\right)\dot{\theta}$

using Lagrange's equation $\qquad \dfrac{d}{dt}\left(\dfrac{\partial L}{\partial \dot{\theta}}\right) - \dfrac{\partial L}{\partial \theta} = 0$

so $\qquad \frac{1}{2}m\left(k^2 + l^2\right)\ddot{\theta} + Mgl\sin\theta = 0$

or $\qquad \ddot{\theta} + \dfrac{gl}{k^2 + l^2}\sin\theta = 0$

when oscillations are small then $\sin\theta = \theta$ and then

$$\ddot{\theta} + \frac{gl}{k^2 + l^2}\theta = 0 \qquad \text{so} \qquad \omega^2 = \sqrt{\frac{gl}{k^2 + l^2}}$$

And time period $\qquad T = \dfrac{2\pi}{\omega} = 2\pi\sqrt{\dfrac{k^2 + l^2}{gl}}$

Or $\qquad\qquad \boxed{T = 2\pi\sqrt{\dfrac{\dfrac{k^2}{l} + l}{g}}}$

8. L-C Circuit

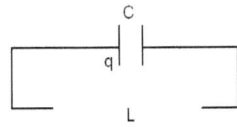

Let us consider an electrical circuit, containing inductance L and Capacitance C.

In this circuit the magnetic energy is analogous to kinetic energy i.e. $\frac{1}{2}Li^2$ corresponds to $\frac{1}{2}mv^2$.

And $i = \dfrac{dq}{dt}$ corresponds to $v = \dfrac{dx}{dt}$. Electrical potential energy of the circuit is $V = \dfrac{q^2}{2C}$

Hence the Lagrangian of the given circuit is $L = T - V$

$$= \frac{1}{2}Li^2 - \frac{q^2}{2C}$$

$$= \frac{1}{2}L\dot{q}^2 - \frac{q^2}{2C}$$

So $\dfrac{\partial L}{\partial \dot{q}} = L\dot{q}$ and $\dfrac{\partial L}{\partial q} = -\dfrac{q}{C}$

Using Lagrange's equation $\dfrac{d}{dt}\left(\dfrac{\partial L}{\partial \dot{q}}\right) - \dfrac{\partial L}{\partial q} = 0$

$$\frac{d}{dt}[L\dot{q}] + \frac{q}{C} = 0 \qquad \text{or} \qquad \ddot{q} + \frac{q}{LC} = 0$$

Comparing it with general equation of SHM $\ddot{q} + \omega^2 q = 0$

$$\omega = \sqrt{\frac{1}{LC}} \quad \text{and the time period of LC oscillation } T = 2\pi\sqrt{LC}$$

And frequency $\boxed{v = \dfrac{1}{2\pi}\sqrt{\dfrac{1}{LC}}}$

9. Radial and Tangential components of any force

Let us consider the motion of a particle of mass m moving in plane with coordinate (r,θ) as generalized coordinate.

And here Cartesian and polar coordinates are related as $x = r\cos\theta$ and $y = r\sin\theta$

So $\dot{x} = \dot{r}\cos\theta - r\dot{\theta}\sin\theta$ and $\dot{y} = \dot{r}\sin\theta + r\dot{\theta}\cos\theta$

And Kinetic energy $T = \dfrac{1}{2}m\left(\dot{x}^2 + \dot{y}^2\right) = \dfrac{1}{2}m\left(\dot{r}^2 + r^2\dot{\theta}^2\right)$

Here generalized coordinates are $q_1 = r$ and $q_2 = \theta$

So $\dfrac{\partial T}{\partial r} = mr\dot{\theta}^2$, $\dfrac{\partial T}{\partial \dot{r}} = m\dot{r}$, $\dfrac{\partial T}{\partial \theta} = 0$ and $\dfrac{\partial T}{\partial \dot{\theta}} = mr^2\dot{\theta}$

Using Lagrange's equations for these two generalized coordinates

$\dfrac{d}{dt}\left(\dfrac{\partial T}{\partial \dot{r}}\right) - \dfrac{\partial T}{\partial r} = G_r$ and $\dfrac{d}{dt}\left(\dfrac{\partial T}{\partial \dot{\theta}}\right) - \dfrac{\partial T}{\partial \theta} = G_\theta$

Or $m\ddot{r} - mr\dot{\theta}^2 = G_r$ and $\dfrac{d}{dt}\left(mr^2\dot{\theta}\right) = G_\theta$

Now using the definition of generalized force

$G_r = \vec{F}\cdot\dfrac{\partial \vec{r}}{\partial r}$ and $G_\theta = \vec{F}\cdot\dfrac{\partial \vec{r}}{\partial \theta}$

But $\vec{r} = r\cos\theta\hat{i} + r\sin\theta\hat{j}$

Hence $\dfrac{\partial \vec{r}}{\partial r} = \cos\theta\hat{i} + \sin\theta\hat{j} = \dfrac{\vec{r}}{r} = \hat{r}$ and $\dfrac{\partial \vec{r}}{\partial \theta} = -r\sin\theta\hat{i} + r\cos\theta\hat{j} = r\hat{\theta}$

So $G_r = \vec{F}.\hat{r} = F_r$　　or　　$F_r = m\ddot{r} - mr\dot{\theta}^2$

And $G_\theta = \vec{F}.r\hat{\theta} = r\vec{F}.\hat{\theta} = rF_\theta$　or　$rF_\theta = \dfrac{d}{dt}\left(mr^2\dot{\theta}\right)$

But as we know that $mr^2\dot{\theta} = mvr = J$ i.e. angular momentum and its derivative w.r.t. time is torque $\left(rF_\theta\right)$

So radial and tangential components are

$F_r = m\left(\ddot{r} - r\dot{\theta}^2\right)$　　　　　　and　　　　　　$F_\theta = m\left(r\ddot{\theta} + 2\dot{r}\dot{\theta}\right)$

10. Two Particle system

Let there are two particles of masses m_1 and m_2 at $\vec{r_1}$ and $\vec{r_2}$ respectively w.r.t. origin with $\vec{R}$ as the position vector of centre of mass O.

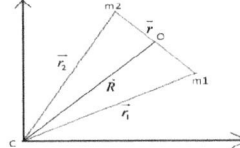

Suppose $\vec{r_1'}$ and $\vec{r_2'}$ are position vectors w.r.t. to O and $\vec{r} = \vec{r_1} - \vec{r_2} = \vec{r_1'} - \vec{r_2'}$. Here kinetic energy of the system $T = \dfrac{1}{2}\left(m_1 + m_2\right)\dot{R}^2 + T'$ where T' is kinetic energy of the system about O.

And $T' = \dfrac{1}{2}m_1\dot{r_1'}^2 + \dfrac{1}{2}m_2\dot{r_2'}^2$

Now $\displaystyle\sum_{i=1}^{N} m_i\vec{r_i'} = 0$　　　or　　　$m_1\vec{r_1'} + m_2\vec{r_2'} = 0$　　　or　　　$\vec{r_2'} = -\dfrac{m_1}{m_2}\vec{r_1'}$

And $\vec{r_1'} = \dfrac{m_2}{m_1 + m_2}\vec{r}$　　similarly　　$\vec{r_2'} = -\dfrac{m_1}{m_1 + m_2}\vec{r}$

Now $T = \dfrac{1}{2}\left(m_1 + m_2\right)\dot{R}^2 + T' = \dfrac{1}{2}\left(m_1 + m_2\right)\dot{R}^2 + \dfrac{1}{2}m_1\dot{r_1'}^2 + \dfrac{1}{2}m_2\dot{r_2'}^2$

$= \dfrac{1}{2}\left(m_1 + m_2\right)\dot{R}^2 + \dfrac{1}{2}m_1\dfrac{m_2^2}{\left(m_1 + m_2\right)^2}\dot{r}^2 + \dfrac{1}{2}m_2\dfrac{m_1^2}{\left(m_1 + m_2\right)^2}\dot{r}^2$

$= \dfrac{1}{2}\left(m_1 + m_2\right)\dot{R}^2 + \dfrac{1}{2}\dfrac{m_1 m_2}{m_1 + m_2}\dot{r}^2$

$= \dfrac{1}{2}M\dot{R}^2 + \dfrac{1}{2}\mu\dot{r}^2$　　　here $M = m_1 + m_2 =$ total mass of the system and

$$\mu = \dfrac{m_1 m_2}{m_1 + m_2} = \text{reduced mass of the system.}$$

18

The Lagrangian of the system $\quad L = T - V = \dfrac{1}{2} M\dot{R}^2 + \dfrac{1}{2} \mu\dot{r}^2 - V(r)$

And Lagrange's equation is $\quad \dfrac{d}{dt}\left(\dfrac{\partial L}{\partial \dot{r}}\right) - \dfrac{\partial L}{\partial r} = 0$

Or $\quad \dfrac{d}{dt}\left(\mu\vec{r}\right) + \dfrac{\partial V(r)}{\partial r} = 0 \qquad$ or $\qquad \boxed{\mu\ddot{r} + \dfrac{\partial V(r)}{\partial r} = 0}$

And $\quad \dfrac{d}{dt}\left(\dfrac{\partial L}{\partial \dot{R}}\right) - \dfrac{\partial L}{\partial R} = 0 \qquad$ or $\qquad \dfrac{d}{dt}\left(M\dot{R}\right) = 0$

Or $\quad M\ddot{\vec{R}} = 0 \qquad\qquad\qquad$ or $\qquad \vec{R} = $ constant

11. Electrical Circuit containing L,C and R in series

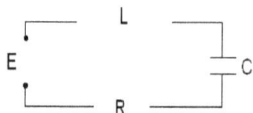

When any circuit involves any inductance L, capacitance C and resistance R then its Lagrangian can be given as $L_E = T_m - V_E$ where T_m is the magnetic energy of the electrical energy of the circuit which corresponds to potential energy of the system.

Suppose q is the charge flowing through the circuit then $T_m = \dfrac{1}{2} Li^2 = \dfrac{1}{2} L\dot{q}^2$

And $V_E = \dfrac{q^2}{2C} - qE$

Hence Lagrangian of this circuit be $L_E = T_m - V_E = \dfrac{1}{2} Li^2 - \left(\dfrac{q^2}{2C} - qE\right)$

Now $\dfrac{\partial L_E}{\partial \dot{q}} = L\dot{q} \qquad$ and $\qquad \dfrac{\partial L_E}{\partial q} = -\left(\dfrac{q}{C} - E\right)$

The dissipative force due to resistance R is Q = -iR

So using Lagrange's equation $\dfrac{d}{dt}\left(\dfrac{\partial L_E}{\partial \dot{q}_j}\right) - \dfrac{\partial L_E}{\partial q_j} = Q_j$

$\dfrac{d}{dt}\left(\dfrac{\partial L_E}{\partial \dot{q}}\right) - \dfrac{\partial L_E}{\partial q} = -iR \qquad$ or $\qquad L\ddot{q} + \dfrac{q}{C} - E = -iR$

Or $\quad \boxed{L\ddot{q} + \dfrac{q}{C} = E - iR}$

Above is the required equation of motion of charge in the said circuit containing L,C and R.

19

Advantages of Lagrangian mechanics over Newtonian mechanics

In Newtonian mechanics, the equation of motion involves many vector quantities e.g. force, momentum, velocity etc., which cannot avoid constraints present in the system. In Lagrangian mechanics only scalars like potential and kinetic energies are only taken care of. In Lagrangian mechanics the form of Lagrange's equations remain invariant w.r.t. any coordinate system. In Newtonian mechanics many forces may not be known due to presence of constraints in the system but in Lagrangian mechanics the value of kinetic and potential energies can easily be calculated. Further Lagrangian mechanics uses generalized coordinate system which do not depend on any particular coordinate system, whereas Newtonian mechanics uses requires symmetric systems form a particular coordinate system. The use of generalized coordinates allows constraints in the system due to which calculating and finding the solution becomes much easier.

Hamiltonian Formulation

In Lagrangian formulation, generalized coordinates $q_1, q_2, \ldots\ldots\ldots\ldots q_n$ are used to find the equation of motion. In Lagrangian formulation, equation of motion are in the form of a set of second order differential equations which are sometimes difficult to solve. Hamilton proposed another formulation known as Hamiltonian dynamics, which uses generalized momenta $p_1, p_2 \ldots\ldots\ldots\ldots p_n$ in place of generalized velocities $\dot{q}_1, \dot{q}_2, \ldots\ldots\ldots\ldots \dot{q}_n$ used in Lagrangian. Hamilton derived a set of first order differential equations instead of a set of second order differential equation. He derived a set of first order differential equations of motion in the form of generalized coordinate q_j and momenta p_j to know the nature of a dynamical system. These equations can easily be solved. These are the main differences between Lagrangian and Hamiltonian equations.

Hamiltonian Function H

Let us consider a generalized Lagrangian l of a system by

$$L = L\left(q_1, q_2 \ldots\ldots\ldots\ldots q_n, \dot{q}_1, \dot{q}_2 \ldots\ldots\ldots\ldots \dot{q}_n, t\right) \quad \text{or} \quad L = L\left(q_k, \dot{q}_k, t\right)$$

Hence
$$\frac{dL}{dt} = \sum_k \frac{\partial L}{\partial q_k} \cdot \frac{dq_k}{dt} + \sum_k \frac{\partial L}{\partial \dot{q}_k} \cdot \frac{d\dot{q}_k}{dt} + \frac{\partial L}{\partial t}$$

From Lagrangian equations $\dfrac{\partial L}{\partial q_k} = \dfrac{d}{dt}\left(\dfrac{\partial L}{\partial \dot{q}_k}\right)$

And
$$\frac{dL}{dt} = \sum_k \frac{d}{dt}\left(\frac{\partial L}{\partial \dot{q}_k}\right) \cdot \dot{q}_k + \sum_k \frac{\partial L}{\partial \dot{q}_k} \cdot \frac{d\dot{q}_k}{dt} + \frac{\partial L}{\partial t}$$

Or
$$\frac{dL}{dt} = \sum_k \frac{d}{dt}\left(\dot{q}_k \frac{\partial L}{\partial \dot{q}_k}\right) + \frac{\partial L}{\partial t}$$

Or
$$\frac{d}{dt}\left(\sum_k \dot{q}_k \frac{\partial L}{\partial \dot{q}_k} - L\right) = -\frac{\partial L}{\partial t}$$

The quantity $\left(\sum_k \dot{q}_k \dfrac{\partial L}{\partial \dot{q}_k} - L \right)$ is sometimes called energy function denoted by h.

So $\qquad h\left(q_1, q_2 ...... q_n, \dot{q}_1, \dot{q}_2 .... \dot{q}_n, t \right) = \sum_k \dot{q}_k \dfrac{\partial L}{\partial \dot{q}_k} - L$

So $\qquad \dfrac{dh}{dt} = -\dfrac{\partial L}{\partial t}$ ....(i)

As L does not depend on time so $\dfrac{\partial L}{\partial t} = 0$ $\qquad$ so $\qquad \dfrac{dh}{dt} = 0$ $\qquad$ or $\qquad$ h=constant.

When Lagrangian is not explicit function of time, the energy function is the constant of motion and this integral is called Jacobi's integral.

But using the above equations

$$\frac{d}{dt}\left(\sum_k p_k \dot{q}_k - L \right) = -\frac{\partial L}{\partial t} \qquad \text{as} \qquad \frac{\partial L}{\partial \dot{q}_k} = p_k \qquad ....(ii)$$

Here the quantity $\left(\sum_k p_k \dot{q}_k - L \right)$ is called the Hamiltonian function H i.e.

$$H = \sum_k p_k \dot{q}_k - L \qquad ....(iii)$$

So from (ii) $\qquad \dfrac{dH}{dt} = 0$ $\qquad$ or $\qquad$ H=constant

So $\qquad H = \sum_k p_k \dot{q}_k - L = \text{constant}$ ....(iv)

As H is constant i.e. conserved, thus equation (iv) is the conservation theorem for the Hamiltonian of the system. Under some conditions, Hamiltonian H is equal to the total energy E of the system.

Conservation of Energy

In a special condition, if the system is conservative i.e. the potential energy V is independent of velocity coordinates $\dot{q}_k$ and the transformation equations for coordinate do not contain time explicitly i.e.

$$\vec{r}_i = \vec{r}_i \left(q_1, q_2, ... q_k, ... q_n \right)$$

For any conservative system $\qquad \dfrac{\partial V}{\partial \dot{q}_k} = 0$

And $p_k = \dfrac{\partial L}{\partial \dot{q}_k} = \dfrac{\partial}{\partial \dot{q}_k}(T-V) = \dfrac{\partial T}{\partial \dot{q}_k}$

So Hamiltonian $\quad H = \sum_k p_k \dot{q}_k - L = \sum_k \dfrac{\partial T}{\partial \dot{q}_k}\dot{q}_k - L$

But $\quad\quad\quad\quad \sum_k \dfrac{\partial T}{\partial \dot{q}_k}\dot{q}_k = 2T$

For a natural conservative system T and V do not contain any explicit time dependence.

So $\quad\quad\quad\quad$ H = 2T – L = 2T – (T-V)

$\quad\quad\quad\quad\quad\quad$ H = T + V = E (constant)

Here Hamiltonian H represents the total energy of the system and is conserved, when system is conservative and T is homogeneous quadratic function.

Formulation of Hamilton's Canonical Equations of motion

As $\quad\quad\quad\quad H = H(p_j, q_j, t)$

And $\quad\quad\quad\quad H = \sum_j p_j \dot{q}_j - L$

so $\quad\quad\quad\quad dH = \sum_j \dot{q}_j dp_j + \sum_j p_j d\dot{q}_j - dL \quad\quad\quad\quad\quad$ ....(i)

again $\quad\quad\quad L = L(q_j, \dot{q}_j, t)$

so $\quad\quad\quad\quad dL = \sum_j \dfrac{\partial L}{\partial q_j}dq_j + \sum_j \dfrac{\partial L}{\partial \dot{q}_j}dq_j + \dfrac{\partial L}{\partial t}dt \quad\quad\quad$ ....(ii)

using (ii) in (i) $\quad dH = \sum_j \dot{q}_j dp_j + \sum_j p_j d\dot{q}_j - \sum_j \dfrac{\partial L}{\partial q_j}dq_j - \sum_j \dfrac{\partial L}{\partial \dot{q}_j}d\dot{q}_j - \dfrac{\partial L}{\partial t}dt \quad$ ....(iii)

again $\dfrac{\partial L}{\partial \dot{q}_j} = p_j \quad\quad$ and $\quad\quad \dfrac{d}{dt}\left(\dfrac{\partial L}{\partial \dot{q}_j}\right) = \dfrac{\partial L}{\partial q_j} \quad$ or $\quad \dot{p}_j = \dfrac{\partial L}{\partial q_j}$

so $\quad\quad\quad\quad dH = \sum_j \dot{q}_j dp_j + \sum_j p_j d\dot{q}_j - \sum_j \dot{p}_j dq_j - \sum_j p_j d\dot{q}_j - \dfrac{\partial L}{\partial t}dt$

or $$dH = \sum_j \dot{q}_j dp_j - \sum_j \dot{p}_j dq_j - \frac{\partial L}{\partial t} dt \qquad\qquad\dots\dots\dots\dots\dots\dots\text{(iv)}$$

but as we know that $H = H(p_j, q_j, t)$

so $$dH = \sum_j \frac{\partial H}{\partial q_j} dq_j + \sum_j \frac{\partial H}{\partial p_j} dp_j + \frac{\partial H}{\partial t} dt \qquad\qquad\dots\dots\dots\dots\dots\dots\text{(v)}$$

equating eq (iv) and (v) and comparing the coefficients, we get

$$-\dot{p}_j = \frac{\partial H}{\partial q_j} \quad \text{or} \quad \boxed{\dot{p}_j = -\frac{\partial H}{\partial q_j}} \qquad\qquad\dots\dots\dots\dots\dots\dots\text{(vi)}$$

$$\boxed{\dot{q}_j = \frac{\partial H}{\partial p_j}} \quad\dots\dots\dots\text{(vii)} \quad \text{and} \quad \boxed{\frac{\partial H}{\partial t} = -\frac{\partial L}{\partial t}} \qquad\dots\dots\dots\dots\dots\dots\text{(viii)}$$

Above eq (vi), (vii) and (viii) are known as Hamiltonian equations of motion.

General discussion – Hamiltonian Function

- Dor a conservative system, the Hamiltonian $H = T + V =$ total energy of the system and constraints are independent of time.
- It is defined as a function of generalized coordinate q_j generalized momenta p_j and time t i.e. $H = H(p_j, q_j, t)$.
- Hamiltonian can be defined as

$$H = \sum_j p_j \dot{q}_j - L(q_j, \dot{q}_j, t) \quad \text{where } p_j = \frac{\partial L}{\partial \dot{q}_j}$$

- Hamiltonian equations of motion can be derived using Lagrangian function L

$$\dot{p}_j = -\frac{\partial H}{\partial q_j}, \ \dot{q}_j = \frac{\partial H}{\partial p_j}, \ \frac{\partial H}{\partial t} = -\frac{\partial L}{\partial t}$$

- The Hamiltonian remains conserved, when Lagrangian of the system not depends on time explicitly.
- For any time dependent constraint, $H \neq E$ but still total energy remains conserved.

Applications of Hamiltonian

1. Harmonic Oscillator

For any harmonic oscillator, kinetic energy $T = \dfrac{1}{2}m\dot{x}^2$ and potential energy $V = \dfrac{1}{2}kx^2$

So Lagrangian $L = T - V = = \dfrac{1}{2}m\dot{x}^2 - \dfrac{1}{2}kx^2$

And generalized momentum $p_x = \dfrac{\partial L}{\partial \dot{x}} = m\dot{x}$ or $\dot{x} = \dfrac{p_x}{m}$

So $T = \dfrac{1}{2}m\dot{x}^2 = \dfrac{p_x^2}{2m}$

So $H = T + V = \dfrac{p_x^2}{2m} + \dfrac{1}{2}kx^2$

Hence Hamiltonian equations are

$\dot{x} = \dfrac{\partial H}{\partial p_x} = \dfrac{p_x}{m}$ or $p_x = m\dot{x}$ ....(i)

And $\dot{p}_x = -\dfrac{\partial H}{\partial x} = -kx$

...........................(ii)

Using (i) in (ii)

$m\ddot{x} = -kx$ or $\boxed{m\ddot{x} + kx = 0}$ which is the required Hamiltonian equation of any harmonic oscillator.

2. Moving Particle in a central force field

We know that all central forces are conservative in nature and for that $F(r) = -\dfrac{\partial V}{\partial r}$ and

using inverse square law $F = -\dfrac{k}{r^2}$. So $V(r) = -\dfrac{k}{r}$ ....(i)

If m be the mass of the particle then Lagrangian of the system can be $L = T - V(r)$

So $L = \dfrac{1}{2}m\left(\dot{r}^2 + r^2\dot{\theta}^2\right) - V(r)$ ....(ii)

Hamiltonian can be written by replacing $\dot{r}$ and $\dot{\theta}$ by p_r and p_θ where

$p_r = \dfrac{\partial L}{\partial \dot{r}} = m\dot{r}$ and $p_\theta = \dfrac{\partial L}{\partial \dot{\theta}} = mr^2\dot{\theta}$

So $\dot{r} = \dfrac{p_r}{m}$ and $\dot{\theta} = \dfrac{p_\theta}{mr^2}$ ....(iii)

So $H = T + V = \dfrac{1}{2m}\left[p_r^2 + \dfrac{p_\theta^2}{r^2}\right] + V(r)$ ....(iv)

And Hamiltonian equations are

$$\dot{q}_k = \frac{\partial H}{\partial p_k} \quad \text{and} \quad \dot{p}_k = -\frac{\partial H}{\partial q_k}$$

Hence $\dot{r} = \dfrac{\partial H}{\partial p_r} = \dfrac{p_r}{m}$ and $\dot{p}_r = -\dfrac{\partial H}{\partial r} = \dfrac{p_\theta^2}{mr^3} - \dfrac{\partial V}{\partial r}$

And $\dot{\theta} = \dfrac{\partial H}{\partial p_\theta} = \dfrac{p_\theta}{mr^2}$ and $\dot{p}_\theta = -\dfrac{\partial H}{\partial \theta} = 0$

So $p_\theta = \text{constant} = mr^2\dot{\theta}$ ....(v)

And $-m\ddot{r} = -\dfrac{p_\theta^2}{mr^3} + \dfrac{\partial V}{\partial r}$ or $m\ddot{r} - \dfrac{p_\theta^2}{mr^3} + \dfrac{\partial V}{\partial r} = 0$
..........................(vi)

And in case of square law $V(r) = -\dfrac{k}{r}$ and $F(r) = -\dfrac{\partial V}{\partial r} = -\dfrac{k}{r^2}$

So eq (vi) reduces to $\boxed{m\ddot{r} - \dfrac{p_\theta^2}{mr^3} + \dfrac{k}{r^2} = 0}$

3. Simple Pendulum

For a simple pendulum, kinetic energy $T = \dfrac{1}{2}ml^2\dot{\theta}^2$ and potential energy $V = mgl(1-\cos\theta)$.

So Lagrangian $L = T - V = \dfrac{1}{2}ml^2\dot{\theta}^2 - mgl(1-\cos\theta)$ ....(i)

So generalized momentum $p_\theta = \dfrac{\partial L}{\partial \dot{\theta}} = ml^2\dot{\theta}$

Or $\dot{\theta} = \dfrac{p_\theta}{ml^2}$

And Hamiltonian $H = \sum_j p_j \dot{q}_j - L = p_\theta \dot{\theta} - L$

$$= \frac{p_\theta^2}{ml^2} - \left[\frac{ml^2 p_\theta^2}{2m^2 l^4} - mgl(1-\cos\theta)\right]$$

$$= \frac{p_\theta^2}{2ml^2} + mgl(1-\cos\theta) \quad \text{..........................(ii)}$$

So Hamiltonian equations are

$$\dot{\theta} = \frac{\partial H}{\partial p_\theta} = \frac{p_\theta}{ml^2} \quad \text{and} \quad \dot{p}_\theta = -\frac{\partial H}{\partial \theta} = -mgl\sin\theta$$

So $p_\theta = ml^2\dot{\theta}$ and $\dot{p}_\theta = -mgl\sin\theta$

Or $\dfrac{d}{dt}(p_\theta) = -mgl\sin\theta$

$\dfrac{d}{dt}(ml^2\dot{\theta}) + mgl\sin\theta = 0$

$$\text{Or} \qquad \ddot{\theta} + \frac{g}{l}\sin\theta = 0$$

For small oscillations $\sin\theta \simeq \theta$

So $$\ddot{\theta} + \frac{g}{l}\theta = 0$$

Above is the required Hamiltonian equation of motion of a simple pendulum.

4. Charged particle moving in an electromagnetic field

When any charged particle of charge q moves in any electromagnetic field then Lagrangian is given by

$$L = T - V = T - q\phi + q(\vec{v}, \vec{A})$$

$$= \frac{1}{2}\sum_k mv_k^2 + q\sum_k v_k A_k - q\phi$$

So $$p_k = \frac{\partial L}{\partial \dot{q}_k} = \frac{\partial L}{\partial v_k} = mv_k + qA_k$$

And $$H = \sum_k p_k \dot{q}_k - L = \sum_k \left(mv_k^2 + qA_k v_k\right) - L$$

$$= mv^2 + q(\vec{v}.\vec{A}) - \frac{1}{2}mv^2 + q\phi - q(\vec{v}.\vec{A})$$

$$= \frac{1}{2}mv^2 + q\phi \qquad\qquad \dots\dots\dots\dots\dots\dots\dots(i)$$

And $$v_k = \frac{p_k}{m} - \frac{q}{m}A_k$$

So $$H = \sum_k \frac{1}{2m}\left(p_k - qA_k\right)^2 + q\phi$$

Or $$H = \frac{1}{2m}\left(\vec{p} - q\vec{A}\right)^2 + q\phi$$

Hence Hamiltonian equations are

$$v_k = \frac{\partial H}{\partial p_x} = \frac{1}{m}\left(p_x - qA_k\right) \qquad \text{or} \qquad \vec{v} - \frac{1}{m}\left(\vec{p} - q\vec{A}\right)$$

$$\dot{p}_k = -\frac{\partial H}{\partial q_k} \qquad\qquad\quad \text{or} \qquad \dot{\vec{p}} = -q\nabla\phi + q\nabla(\vec{v}.\vec{A})$$

Above two are required Hamiltonian equations.

5. Compound Pendulum

For any compound pendulum, Lagrangian is given by

$$L = T - V = \frac{1}{2}I\dot{\theta}^2 + Mgl\cos\theta$$

And generalized momentum $p_\theta = \dfrac{\partial L}{\partial \dot{\theta}} = I\dot{\theta}$

Hence Hamiltonian is given by $H = \displaystyle\sum_k p_k \dot{q}_k - L$

So $\quad H = p_\theta \dot{\theta} - \dfrac{1}{2} I \dot{\theta}^2 - Mgl \cos\theta$

Or $\quad H = \dfrac{p_\theta^2}{2I} - Mgl \cos\theta$

So Hamiltonian equations are given as

$\dot{\theta} = \dfrac{\partial H}{\partial p_\theta} \qquad\qquad$ or $\qquad\qquad \dot{\theta} = \dfrac{p_\theta}{I}$

And $\dot{p}_\theta = -\dfrac{\partial H}{\partial \theta} \qquad$ or $\qquad\qquad \dot{p}_\theta = -Mgl \sin\theta$

So $\quad I\ddot{\theta} = -Mgl \sin\theta \qquad$ and for small θ, sin θ = θ

So $\quad \ddot{\theta} + \dfrac{Mgl}{I}\theta = 0$

And time period of the compound pendulum is given by $T = 2\pi\sqrt{\dfrac{I}{Mgl}}$

This is the required Hamiltonian equation.

6. Two dimensional harmonic oscillator

(A) In Cartesian Coordinates

Here $L = T - V = \dfrac{1}{2}m\left(\dot{x}^2 + \dot{y}^2\right) - \dfrac{1}{2}k\left(x^2 + y^2\right)$

So $p_x = \dfrac{\partial L}{\partial \dot{x}} = m\dot{x} \qquad$ or $\qquad \dot{x} = \dfrac{p_x}{m}$ and $\dot{y} = \dfrac{p_y}{m}$

So Hamiltonian $H = \displaystyle\sum_k p_k \dot{q}_k - L$

$= p_x \dot{x} + p_y \dot{y} - \dfrac{1}{2}m\left(\dot{x}^2 + \dot{y}^2\right) + \dfrac{1}{2}k\left(x^2 + y^2\right)$

$= \dfrac{1}{2m}\left(p_x^2 + p_y^2\right) + \dfrac{1}{2}k\left(x^2 + y^2\right)$

So $\dot{x} = \dfrac{\partial H}{\partial p_x}$ or $\dot{x} = \dfrac{p_x}{m} \qquad$ and $\qquad \dot{p}_x = -\dfrac{\partial H}{\partial x} = -kx$

And $\dot{y} = \dfrac{\partial H}{\partial p_y}$ or $\dot{y} = \dfrac{p_y}{m} \qquad$ and $\qquad \dot{p}_y = -\dfrac{\partial H}{\partial y} = -ky$

So Hamiltonian equations become

$m\ddot{x} + kx = 0 \qquad\qquad$ and $\qquad m\ddot{y} + ky = 0$

(B) Polar Coordinates

Here $x = r\cos\theta$ and $y = r\sin\theta$

And Lagrangian L $= T - V = \dfrac{1}{2}m\left(\dot{x}^2 + \dot{y}^2\right) - \dfrac{1}{2}k\left(x^2 + y^2\right)$

$$= \dfrac{1}{2}m\left(\dot{r}^2 + r^2\dot{\theta}^2\right) - \dfrac{1}{2}kr^2$$

So $p_r = \dfrac{\partial L}{\partial \dot{r}} = m\dot{r}$ or $\dot{r} = \dfrac{p_r}{m}$

And $p_\theta = \dfrac{\partial L}{\partial \dot{\theta}} = mr^2\dot{\theta}$ or $\dot{\theta} = \dfrac{p_\theta}{mr^2}$

Now $H = \sum\limits_k p_k\dot{q}_k - L = p_r\dot{r} + p_\theta\dot{\theta} - \dfrac{1}{2}m\left(\dot{r}^2 + r^2\dot{\theta}^2\right) + \dfrac{1}{2}kr^2$

Or $H = \dfrac{1}{2m}\left(p_r^2 + \dfrac{p_\theta^2}{r^2}\right) + \dfrac{1}{2}kr^2$

So Hamiltonian equations are

$\dot{r} = \dfrac{\partial H}{\partial p_r} = \dfrac{p_r}{m}$ and $\dot{p}_r = -\dfrac{\partial H}{\partial r} = \dfrac{p_\theta^2}{mr^3} - kr$

$\dot{\theta} = \dfrac{\partial H}{\partial p_\theta} = \dfrac{p_\theta}{mr^2}$ and $\dot{p}_\theta = -\dfrac{\partial H}{\partial \theta} = 0$

So $p_\theta = mr^2\dot{\theta} = $ constant   and $\dot{p}_r = m\ddot{r} = \dfrac{p_\theta^2}{mr^3} - kr$

Or $m\ddot{r} - mr\dot{\theta}^2 + kr = 0$

The above is the Hamiltonian equations for two dimensional harmonic oscillators.

7. Atwood's Machine

For this problem the constraint is $x_1 + x_2 = l$

Differentiating the above we get $\dot{x}_1 = -\dot{x}_2$

So kinetic energy $T = \dfrac{1}{2}m_1\dot{x}_1^2 + \dfrac{1}{2}m_2\dot{x}_2^2 = \dfrac{1}{2}\left(m_1 + m_2\right)\dot{x}_1^2$

And potential energy $V = -m_1 gx_1 - m_2 g\left(l - x_1\right)$

So Lagrangian L $= T - V = \dfrac{1}{2}\left(m_1 + m_2\right)\dot{x}_1^2 + m_1 gx_1 - m_2 gx_1 + m_2 gl$

$$= \frac{1}{2}\left(m_1 + m_2\right)\dot{x}_1^{\ 2} + \left(m_1 - m_2\right)gx_1 + V_0$$

Where $V_0 = m_2 gl$

So generalized momentum $p_{x_1} = \dfrac{\partial L}{\partial \dot{x}_1} = \left(m_1 + m_2\right)\dot{x}_1$

And Hamiltonian $H = \sum_k p_k \dot{q}_k - L = p_{x_1}\dot{x}_1 - L$

$$= \left(m_1 + m_2\right)\dot{x}_1^{\ 2} - \frac{1}{2}\left(m_1 + m_2\right)\dot{x}_1^{\ 2} - \left(m_1 - m_2\right)gx_1 - V_0$$

$$= \frac{1}{2}\left(m_1 + m_2\right)\dot{x}_1^{\ 2} - \left(m_1 - m_2\right)gx_1 - V_0$$

$$= \frac{p_{x_1}^{\ 2}}{2\left(m_1 + m_2\right)} - \left(m_1 - m_2\right)gx_1 - V_0$$

So Hamilton's equations are

$$\dot{p}_{x_1} = -\frac{\partial H}{\partial x_1} = \left(m_1 - m_2\right)g \qquad \text{and} \qquad \dot{x}_1 = \frac{\partial H}{\partial p_{x_1}} = \frac{p_{x_1}}{\left(m_1 + m_2\right)}$$

Or $\qquad p_{x_1} = \left(m_1 + m_2\right)\dot{x}_1 \qquad$ or $\qquad \dot{p}_{x_1} = \left(m_1 + m_2\right)\ddot{x}_1$

So we get $\qquad \left(m_1 - m_2\right)g = \left(m_1 + m_2\right)\ddot{x}_1$

Or $\qquad \boxed{\ddot{x}_1 = \frac{\left(m_1 - m_2\right)}{\left(m_1 + m_2\right)}g}$

Above is the required Hamiltonian equation of motion for the mentioned problem.

Introduction to Quantum Mechanics

We know that Newtonian mechanics deals with the motion of particles under the applications of external forces, and quantities like position of particle, mass, velocity and acceleration of motion can easily be found. This particular assumption is valid as far as our everyday life is concerned because things are macro in size and Newtonian mechanics provides the correct explanation of these macro movements, where observed quantities are in good agreement with the expected values. In classical mechanics, determination of position and momentum of any system which is macroscopic can be found with same accuracy by using Newton's law of motion provided their initial position and momentum are known.

Quantum mechanics also deals with relationships between observable magnitudes but observable magnitudes changes its definition as per uncertainty principle according to which the position and momentum of particle cannot be accurately measured simultaneously, whereas in Newtonian mechanics both physical quantities are assumed to have definite ascertainable values at every instant. In quantum mechanics quantities are measured in terms of probabilities instead of ascerting the values. In quantum mechanics, when a micro particle is in motion, its position can be anywhere within the wave packet, which extends throughout a region of space, which make the description of the particle meaningless.

Hence quantum mechanics describes the physical quantities in probabilities by considering number of allowed probabilities. This transition from deterministic classical physics to probabilistic quantum physics is best expressed in terms of a function, called wave function.

Wave function

We know that a wave is always associated with the moving quantum mechanical particle, known as matter waves, and the quantities whose variations make up matter waves is called the wave function.

The wave function is represent by a Greek letter ψ (psi) which consists of real and imaginary part like

$$\psi = A + iB$$

The wave function is a mathematical function representing the space-time behavior of each quantum mechanical particle. It is single valued and finite. Wave function is square Integrable function. Its magnitude is large when probability of finding the particle is high and is small in the regions where probability is low. Hence, the wave function ψ measures the probability of the particle around a particular position.

Probability and Probability density

As wave function ψ is given by $\psi = A + iB$

Conjugate of which $\psi^* = A - iB$

So that $\psi\psi^* = A^2 + B^2 = |\psi|^2$

The value $|\psi|^2$ represents a particular place at a particular time and is proportional to the probability of finding the particle there at that time.

So probability density $= |\psi|^2 = \psi\psi^*$

Probability density can be defined as the probability of finding the particle per unit volume of a given space at a particular time.

The probability of finding the particle in the volume element dV at (x,y,z and t) $= \psi\psi^* \, dV$

The probability of finding the particle anywhere in space can be written as

$$P = \int\limits_{x_1}^{x_2} \int\limits_{y_1}^{y_2} \int\limits_{z_1}^{z_2} \psi\psi^* \, dxdydz$$

In any one dimensional case the probability of locating the particle within x direction

$$P = \int\limits_{x_1}^{x_2} \psi\psi^* \, dx$$

Requirements for a physically acceptable wave function

(i) ψ must be continuous and single valued everywhere in space.

(ii) $\dfrac{\partial \psi}{\partial x}, \dfrac{\partial \psi}{\partial y}, \dfrac{\partial \psi}{\partial z}$ must be continuous and single valued everywhere in space.

(iii) ψ must follow the condition of Normalization, which ensures the presence of the particle in space and total probability becomes unity.

i.e. a wave function is said to be Normalized when

$$P = \int\limits_{-\infty}^{\infty} \psi^* \psi \, dx = 1 \ \text{(in x-direction)}$$

And $\ P = \int\limits_{-\infty}^{\infty} \int\limits_{\infty}^{\infty} \int\limits_{\infty}^{\infty} \psi^* \psi \, dxdydz = 1$ (for three dimensional space)

P=1 represents that quantum mechanical particle exists definitely somewhere in space of consideration.

Orthogonal wave function

If there exists two wave functions ψ_a and ψ_b such that

$$\int \psi_a^* \psi_b \, dV = \int \psi_a \psi_b^* \, dV = 0 \ \text{When} \ a \neq b$$

Here the integral is taken over the whole space. If both functions are simultaneously normalized then

$$\int \psi_a^* \psi_a \, dV = \int \psi_b \psi_b^* \, dV = 1$$

The set of wave function which are both normalized and are also orthogonal are called ortho-normal wave functions.

So $\qquad \int \psi_a^* \psi_b \, dV = 0 \ \text{if} \ a \neq b$

$$=1 \qquad \text{if } a=b$$

If we have a complete set of orthogonal wave functions, then any other wave function can also be expended as a set of orthogonal wave functions if that gives solution to Schrödinger wave equation.

Eigen values and Eigen functions

A physical system is characterized by its position, momentum, energy etc. In case a wave function corresponding to a system is assigned, the state of the system can also be precisely known. In case of any change in the system's state, the associated wave function also changes accordingly, if an operator operating on a function will produce the same function multiplied by a constant factor, the function is called the Eigen function and the constant is known as the Eigen value of the given operator.

If any operator $\hat{O}$ operates on any function $f_n(x)$ and gives a set of wave functions $f_n(x)$ multiplied by a general set of Eigen values C_n which may be defined by the equation

$$\hat{O}f_n(x) = C_n f_n(x)$$

Expectation values

We know that any dynamical variable is defined as a variable which depends on coordinates (x,y,z,t). for example position, potential energy, kinetic energy, momentum etc. these dynamic quantities forms the basis of physical measurements and are known as observables. In quantum mechanics every observable is associated or represented by an operator which acts on a wave function ψ to give a new wave function. The expectation value or average value of a dynamical quantity is the mathematical expectation of the result of a single measurement. To correlate experimental and theoretical results, expectation value of any parameter is defined as

$$<x> = \frac{\int_{-\infty}^{\infty} x|\psi|^2\, dx}{\int_{-\infty}^{\infty} |\psi|^2\, dx} = \frac{\int_{-\infty}^{\infty} \psi^*\psi\, x\, dx}{\int_{-\infty}^{\infty} \psi^*\psi\, dx}$$

If the function is normalized wave function then

$$<x> = \int_{-\infty}^{\infty} x|\psi|^2\, dx$$

Hence average or expectation value of any function $f(x)$ is given by

$$<f(x)> = \frac{\int_{-\infty}^{\infty} \psi^*\psi\, f(x)\, dx}{\int_{-\infty}^{\infty} \psi^*\psi\, dx}$$

Basic postulates of quantum mechanics

1. A wave function can be used to identify the space time behavior of a particle in a physical system. This function and its space derivatives are continuous, single valued and finite.

2. in quantum mechanics, each physical parameter is associated with quantum mechanical operator.

3. wave function is used to find the probability of finding the particle anywhere in space parameters with $P = \int\limits_{-\infty}^{\infty} \psi^{*}\psi\, dx = 1$

4. Expectation value of any dynamical physical quantity is given by

$$< f(x) >= \frac{\int\limits_{-\infty}^{\infty} \psi^{*}\psi f(x) dx}{\int\limits_{-\infty}^{\infty} \psi^{*}\psi\, dx}$$

5. Eigen value of any dynamical variable is given by $\hat{O} f_n(x) = C_n f_n(x)$

Schrödinger Time Dependent Wave Equation

Schrödinger in 1926 proposed a wave function as a development of de-Broglie hypothesis of the wave properties of matter. Schrödinger wave equation is the fundamental equation of wave mechanics. This is the differential equation of the de-Broglie waves associated with particles and describes the motion of particles.

According to de-Broglie hypothesis, wave function of a particle moving freely in the x-direction is given by

$$\psi = Ae^{-i\omega\left(t-\frac{x}{v}\right)}$$

But $\omega = 2\pi v$ and $v = v\lambda$ so $\psi = Ae^{-2\pi i\left(vt-\frac{x}{\lambda}\right)}$

Energy $E = hv = 2\pi\hbar v$, $\lambda = \dfrac{h}{p} = \dfrac{2\pi\hbar}{p}$, where $\hbar = \dfrac{h}{2\pi}$ modified Planck's constant.

So free particle wave equation reduces to

$$\psi = Ae^{-\frac{i}{\hbar}\left(Et-px\right)}$$

.................(i)

Differentiating eq (i) with respect to 't'

$$\frac{d\psi}{dt} = -\frac{iE}{\hbar} Ae^{-\frac{i}{\hbar}\left(Et-px\right)}$$

$$= -\frac{iE}{\hbar}\psi$$

So energy operator $\boxed{E = i\hbar\frac{\partial}{\partial t}}$ ....(ii)

And differentiating eq (i) w.r.t. 'x'

$$\frac{d\psi}{dx} = \frac{ip}{\hbar} Ae^{-\frac{i}{\hbar}\left(Et-px\right)}$$

$$= \frac{ip}{\hbar}\psi$$

So momentum operator $\boxed{p = \frac{\hbar}{i}\frac{\partial}{\partial x}}$ ....(iii).

As the total energy of th eparticle

$$E = K.E. + P.E.$$

$$E = \frac{p^2}{2m} + V$$ ....(iv)

$$E\psi = \left(\frac{p^2}{2m}\right)\psi + V\psi$$

Using (ii) and (iii) in above

$$i\hbar\frac{\partial\psi}{\partial t} = \left(\frac{\hbar}{i}\frac{\partial}{\partial x}\right)^2 \frac{1}{2m}\psi + V\psi$$

$$\boxed{i\hbar\frac{\partial\psi}{\partial t} = -\frac{\hbar^2}{2m}\frac{\partial^2\psi}{\partial x^2} + V\psi}$$ ....(v)

Equation (v) represents Scroedinger time dependent wave equation in +x direction.

For 3-D space

$$\boxed{i\hbar\frac{\partial\psi}{\partial t} = -\frac{\hbar^2}{2m}\left(\frac{\partial^2\psi}{\partial x^2} + \frac{\partial^2\psi}{\partial y^2} + \frac{\partial^2\psi}{\partial z^2}\right) + V\psi}$$ ....(vi)

Schrödinger Time -independent Wave Equation

Time dependent Schrödinger equation governs by the wave function ψ which is a function of both position and time, here potential energy of a moving particle is also a function of position and time. In many practical problems potential energy need not to be dependent on time and there we cannot apply Schrödinger time dependent wave equation and its time independent form is to be used as below.

We know that de-Broglie wave equation for a free particle is given by

$$\psi = Ae^{-\frac{i}{\hbar}\left(Et-px\right)} \qquad \text{(here } \psi \text{ is function of x and t)}$$

$$= Ae^{-\frac{i}{\hbar}Et}\, e^{\frac{i}{\hbar}px}$$

Or $\qquad \psi = \psi_0 e^{-\frac{i}{\hbar}Et}$ ....(i)

Where $\psi_0 = Ae^{\frac{i}{\hbar}px}$ is time independent wave function

Now differentiating (i) w.r.t. 't'

$$\frac{\partial \psi}{\partial t} = -\frac{iE}{\hbar}\psi_0 e^{-\frac{i}{\hbar}Et} \qquad\qquad ....(ii)$$

And differentiating (i) twice w.r.t. 'x'

$$\frac{\partial^2 \psi}{\partial x^2} = \frac{\partial^2 \psi_0}{\partial x^2}e^{-\frac{i}{\hbar}Et} \qquad\qquad ....(iii)$$

Using (i), (ii) and (iii) in Schrödinger time dependent wave equation $i\hbar\frac{\partial \psi}{\partial t} = -\frac{\hbar^2}{2m}\frac{\partial^2 \psi}{\partial x^2}+V\psi$, we get

$$i\hbar\left(-\frac{iE}{\hbar}\right)\psi_0 e^{-\frac{i}{\hbar}Et} = -\frac{\hbar^2}{2m}\frac{\partial^2 \psi_0}{\partial x^2}e^{-\frac{i}{\hbar}Et}+V\psi_0 e^{-\frac{i}{\hbar}Et}$$

Or $\qquad E\psi_0 = -\frac{\hbar^2}{2m}\frac{\partial^2 \psi_0}{\partial x^2}+V\psi_0$

Or $\qquad \boxed{\dfrac{\partial^2 \psi_0}{\partial x^2}+\dfrac{2m}{\hbar^2}(E-V)\psi_0 = 0}$ ....(iv)

This equation (iv) represents Schrödinger time independent wave equation in one-dimension.

In 3-dimensional space

$$\nabla^2 \psi_0 + \frac{2m}{\hbar^2}(E - V)\psi_0 = 0$$

.................(v)

For any free particle V = 0 and eq (v) reduces to

$$\nabla^2 \psi_0 + \frac{2m}{\hbar^2} E\psi_0 = 0$$

.................(vi)

Applications of Schrödinger Time Independent Wave Equation

(A) Particle in a one-dimensional box (Infinite Square well potential)

Let us consider a particle moving inside a box along x-direction and particle is bouncing back and forth between the walls of the box of width L. the both sides of the box are restricted using infinite potential V.

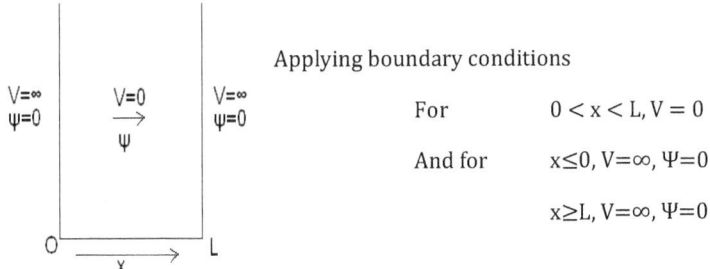

Applying boundary conditions

For $0 < x < L, V = 0$

And for $x \leq 0, V = \infty, \Psi = 0$

              $x \geq L, V = \infty, \Psi = 0$

hence particle cannot exist outside the box so for wave function Ψ is zero for $x \leq 0$ and $x \geq L$ but for $0 < x < L$, Ψ is governed and produced using Schrödinger time independent wave equation with V=0.

So $\frac{\partial^2 \psi}{\partial x^2} + \frac{2m}{\hbar^2} E\psi = 0$ ....(i)

$\frac{\partial^2 \psi}{\partial x^2} + k^2\psi = 0$ where $k^2 = \frac{2mE}{\hbar^2}$ is any constant. ....(ii)

The general solution of (ii) is

$$\psi = A\sin kx + B\cos kx$$

..............(iii)

Applying boundary conditions for (iii)

At x =0, Ψ=0

$$0 = A \sin 0 + B$$

So B =0

At x = L, Ψ=0

$$0 = A \sin kL$$

But $A \neq 0$ so sin kL =0 or $kL = \pm n\pi$

So $k = \pm\dfrac{n\pi}{L}$ ....(iv)

So from (iii) $\psi = A\sin\dfrac{n\pi}{L}x$ ....(v)

Because certainly particle will exist between the range 0 <x <L so using Normalization condition on eq (v)

$$\int_{-\infty}^{\infty} |\psi|^2\, dx = 1$$

Or $\displaystyle\int_0^L A^2 \sin^2\left(\dfrac{n\pi x}{L}\right) dx = 1$

Or $\dfrac{A^2}{2}\displaystyle\int_0^L\left[1-\cos\left(\dfrac{2n\pi x}{L}\right)\right] dx = 1$

Or $\dfrac{A^2}{2}\left[\displaystyle\int_0^L dx - \int_0^L \cos\dfrac{2n\pi x}{L} dx\right] = 1$

Or $\dfrac{A^2}{2}L = 1$ or $A = \sqrt{\dfrac{2}{L}}$ ....(vi)

Using (vi) in (v)

$$\boxed{\psi = \sqrt{\dfrac{2}{L}}\sin\left(\dfrac{n\pi x}{L}\right)}$$ where n = 1,2,3.........   ....(vii)

Now from $k^2 = \dfrac{2mE}{\hbar}$

$$E = \frac{k^2 \hbar^2}{2m} \qquad \text{where } k = \pm\frac{n\pi}{L}$$

So $\boxed{E = \dfrac{n^2 \pi^2 \hbar^2}{2mL^2}}$

Or $\boxed{E = \dfrac{n^2 h^2}{8mL^2}}$ $\qquad\qquad$ where n = 1,2,3.........

So inside an infinitely deep potential well, the particle cannot have an arbitrary energy, but can have only certain discrete energy corresponding to n=1,2,3..... each permitted energy is called Eigen value of the particle and constitute the energy level of the system. The wave function Ψ corresponding to each Eigen value are called Eigen functions.

Although wave function Ψ can be negative or positive but $|\psi|^2$ is always positive. The value of Ψ at x is equal to the probability density of finding that particle there, because Ψ is normalized.

Particle in a three dimensional box

Let us consider a particle being trapped inside a rectangular box of sides a,b,c with its edges along x,y and z axis.

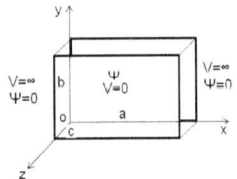

The particle can move freely within the region 0<x<a, 0<y<b and 0<z<c and inside the box its motion is governed by wave function Ψ represented by Schrödinger time independent wave equation in 3-dimensional space with potential inside the box as zero (V=0).

So boundary conditions for this problem can be summarized as

$$\left. \begin{array}{l} 0 < x < a \\ 0 < y < b \\ 0 < z < c \end{array} \right\} \begin{array}{l} V = 0 \\ \\ \psi \end{array}$$

And

$$\left.\begin{array}{ll} x \leq 0, & x \geq a \\ y \leq 0, & y \geq b \\ z \leq 0, & z \geq c \end{array}\right\} \begin{array}{l} V = \infty \\ \psi = 0 \end{array}$$

Schrödinger time independent wave equation for this case can be written as

$$\frac{\partial^2 \psi}{\partial x^2} + \frac{\partial^2 \psi}{\partial y^2} + \frac{\partial^2 \psi}{\partial z^2} + \frac{2m}{\hbar^2} E\psi = 0$$

Or $\quad \left(\frac{\partial^2}{\partial x^2} + \frac{\partial^2}{\partial y^2} + \frac{\partial^2}{\partial z^2} \right) \psi + \frac{2m}{\hbar^2} E\psi = 0 \qquad\qquad$ ....(i)

Here wave function Ψ is a function in three different variable x.y and z. hence by using theorem of separation of variables.

$$\psi_{(x,y,z)} = X_x Y_y Z_z = XYZ \text{ (assume)} \qquad\qquad ....(ii)$$

Using (ii) in (i)

$$\left(\frac{\partial^2}{\partial x^2} + \frac{\partial^2}{\partial y^2} + \frac{\partial^2}{\partial z^2} \right) XYZ + \frac{2m}{\hbar^2} EXYZ = 0$$

$$YZ\frac{\partial^2 X}{\partial x^2} + XZ\frac{\partial^2 Y}{\partial y^2} + XY\frac{\partial^2 Z}{\partial z^2} + \frac{2m}{\hbar^2} EXYZ = 0$$

Dividing above with XYZ

$$\frac{1}{X}\frac{\partial^2 X}{\partial x^2} + \frac{1}{Y}\frac{\partial^2 Y}{\partial y^2} + \frac{1}{Z}\frac{\partial^2 Z}{\partial z^2} + \frac{2m}{\hbar^2} E = 0$$

Or $\quad \frac{1}{X}\frac{\partial^2 X}{\partial x^2} + \frac{1}{Y}\frac{\partial^2 Y}{\partial y^2} + \frac{1}{Z}\frac{\partial^2 Z}{\partial z^2} = -\frac{2m}{\hbar^2} E \qquad\qquad$ ....(iii)

By observing (iii) it is clear that left hand side is a constant, making it clear that right hand side also should be individually equal to some constant.

Let us assume

$$\frac{1}{X}\frac{\partial^2 X}{\partial x^2} = -k_x^2, \quad \frac{1}{Y}\frac{\partial^2 Y}{\partial x^2} = -k_y^2 \text{ and } \quad \frac{1}{Z}\frac{\partial^2 Z}{\partial x^2} = -k_z^2 \qquad$ ....(iv)

So from (iii) $-k_x^2 - k_y^2 - k_z^2 = -\dfrac{2m}{\hbar^2} E$

Or $\qquad\qquad k_x^2 + k_y^2 + k_z^2 = \dfrac{2m}{\hbar^2} E$ ....(v)

And from (iv) $\qquad\qquad \dfrac{1}{X}\dfrac{\partial^2 X}{\partial x^2} = -k_x^2$

Or $\qquad \dfrac{\partial^2 X}{\partial x^2} + k_x^2 X = 0$ Similarly $\quad \dfrac{\partial^2 Y}{\partial y^2} + k_y^2 Y = 0$ and $\quad \dfrac{\partial^2 Z}{\partial z^2} + k_z^2 Z = 0$ ....(vi)

Using the boundary condition on (vi), the general solution of that can be written as

$\qquad\qquad X = A\sin k_x x + B\cos k_x x$ ....(vii)

At x = 0, X = 0 gives B = 0

At x = a, X = 0 given $\sin k_x x = 0 \qquad$ or $\qquad k_x a = n_x \pi$

$\qquad\qquad\qquad\qquad\qquad$ Or $\qquad k_x = \dfrac{n_x \pi}{a}$ ....(viii)

So $\qquad X = A\sin\dfrac{n_x \pi}{a} x$ ....(ix)

Applying Normalization condition on (ix) between 0 to a so that

$\qquad \int_0^a |\psi|^2 dx = 1 \qquad$ or $\qquad A^2 \int_0^a \left|\sin\dfrac{n_x \pi x}{a}\right|^2 dx = 1$

$\qquad A^2 \dfrac{a}{2} = 1 \qquad$ or $\qquad A = \sqrt{\dfrac{2}{a}}$

Using (x) in (ix)

$\qquad\qquad X = \sqrt{\dfrac{2}{a}}\sin\dfrac{n_x \pi}{a} x$ ....(xi-i)

similarly $\qquad Y = \sqrt{\dfrac{2}{b}}\sin\dfrac{n_y \pi}{b} y$ ....(xi-ii) and $\qquad Z = \sqrt{\dfrac{2}{c}}\sin\dfrac{n_z\pi}{c}z$ $\qquad\qquad$ ....(xi-iii)

using (xi-i, ii&iii) in (ii)

$\qquad \psi = XYZ$

$$= \sqrt{\frac{2}{a}}\sin\frac{n_x\pi}{a}x\sqrt{\frac{2}{b}}\sin\frac{n_y\pi}{b}y\sqrt{\frac{2}{c}}\sin\frac{n_z\pi}{c}z$$

$$= \sqrt{\frac{2}{a}\frac{2}{b}\frac{2}{c}}\sin\frac{n_x\pi}{a}x\sin\frac{n_y\pi}{b}y\sin\frac{n_z\pi}{c}z \qquad\qquad ....(xii)$$

Eq (xii) represents the final wave function for a particle trapped inside a three dimensional box of side a,b, and c.

Similarly from (viii) $k_y = \dfrac{n_y\pi}{b}$ $\qquad$ and $\qquad k_z = \dfrac{n_z\pi}{c}$

So from (v) $\qquad \dfrac{n_x^2\pi^2}{a^2}+\dfrac{n_y^2\pi^2}{b^2}+\dfrac{n_z^2\pi^2}{c^2}=\dfrac{2mE_n}{\hbar^2}$

Or $\qquad\qquad E = \dfrac{\hbar^2\pi^2}{2m}\left[\dfrac{n_x^2}{a^2}+\dfrac{n_y^2}{b^2}+\dfrac{n_z^2}{c^2}\right]$ $\qquad\qquad$ ....(xiii)

Eq (xiii) represents the total energy of the system with $n_x, n_y, n_z = 1,2,3$...........

Energy Degeneracy

We know that the energy of a particle in 3-D space is given by

$$E = \frac{\hbar^2\pi^2}{2m}\left[\frac{n_x^2}{a^2}+\frac{n_y^2}{b^2}+\frac{n_z^2}{c^2}\right]$$

For simplicity of calculations, let us assume the box to be a cube of side a=b=c so

$$E = \frac{h^2}{8ma^2}\left(n_x^2 + n_y^2 + n_z^2\right) \qquad\qquad ....(xiv)$$

The energy eigen values are discrete, determined by a set of three quantum numbers n_x, n_y, n_z. values of these should not be zero otherwise wave function reduces to zero.

Let us suppose $\qquad n_x^2 + n_y^2 + n_z^2 = \gamma^2$

So that $\qquad E = \dfrac{h^2}{8ma^2}\gamma^2$ $\qquad\qquad$ ....(xv)

For $n_x = n_y = n_z = 1$, $\gamma^2 = 3$ and $\qquad E = \dfrac{h^2}{8ma^2}.3 = 3E_1 \text{ (let)}$

And the related wave function is

$$\psi = = \dfrac{2\sqrt{2}}{a^{\frac{3}{2}}} \sin\dfrac{n_x\pi}{a}x \sin\dfrac{n_y\pi}{b}y \sin\dfrac{n_z\pi}{c}z \qquad\qquad ....(xvi)$$

Suppose for $n_x = n_y = n_z = 1$, wave function is ψ_{111}.

So by changing the values of n_x, n_y and n_z for higher energy levels in different axial directions, we can see that there is a situation when more than one wave function corresponds to single state. This phenomenon of having one single energy level for than one wave function is known as degeneracy and is is said that that energy level has degenerated itself for those wave functions. The number of times, the energy level is degenerated, it is called g-folded degeneracy.

Let us find different sets of the wave functions for different sets of n_x, n_y and n_z.

n_x	n_y	n_z	γ^2	E	Ψ_{n_x,n_y,n_z}
1	1	1	3	$3E_1$	Ψ_{111}
2	1	1	6	$6E_1$	Ψ_{211}
1	2	1	6	$6E_1$	Ψ_{121}
1	1	2	6	$6E_1$	Ψ_{112}
2	2	1	9	$9E_1$	Ψ_{221}
2	1	2	9	$9E_1$	Ψ_{212}
1	2	2	9	$9E_1$	Ψ_{122}

We can see that for $6E_1$ energy level, there are three wave functions associated with it, resulting in three folded degeneracy of $6E_1$ energy level.

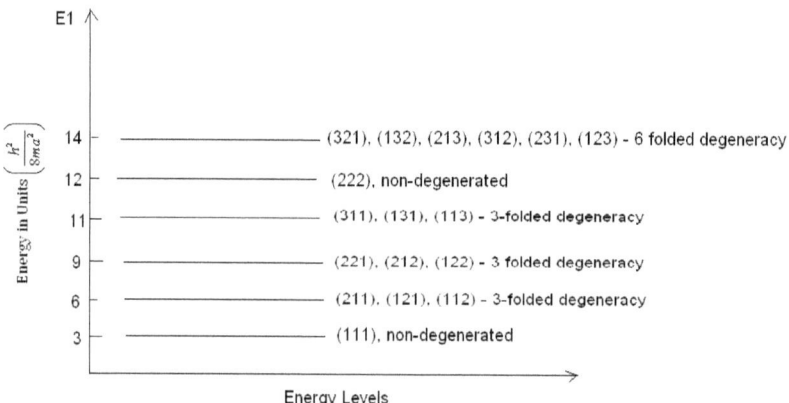

Solved Examples

1. Write down the Lagrange's equation of motion for a particle of mass m falling freely under gravity near the surface of earth.   (Rohilkhand 1997)

Sol.

If x and y axes are in the place of earth surface and z axis vertical to the upward, then kinetic energy of freely falling particle of mass m is

$$T = \frac{1}{2}m\left(\dot{x}^2 + \dot{y}^2 + \dot{z}^2\right)$$

And potential energy V = mgz

So Lagrangian L = T − V = $\frac{1}{2}m\left(\dot{x}^2 + \dot{y}^2 + \dot{z}^2\right) - mgz$

And Lagrange's equations are

$$\frac{d}{dt}\left(\frac{\partial L}{\partial \dot{q}_k}\right) - \frac{\partial L}{\partial q_k} = 0$$

For $q_k = x, y, z$

$$\frac{\partial L}{\partial \dot{x}} = m\dot{x} \; ; \qquad \frac{\partial L}{\partial \dot{y}} = m\dot{y} \; ; \qquad \frac{\partial L}{\partial \dot{z}} = m\dot{z}$$

$$\frac{\partial L}{\partial x} = \frac{\partial L}{\partial y} = 0 \text{ and } \frac{\partial L}{\partial z} = mg$$

So $\quad \dfrac{d}{dt}(m\dot{x}) = 0 \quad ; \qquad \dfrac{d}{dt}(m\dot{y}) = 0 \quad$ and $\quad \dfrac{d}{dt}(m\dot{z}) + mg = 0$

Or $\quad \ddot{x} = 0 \qquad ; \qquad \ddot{y} = 0 \qquad$ and $\quad \ddot{z} + g = 0$

2. A point mass moves in a vertical plane along a given curve in a gravitational field. The equation of motion in parametric form is $x = x(s)$, $z = z(s)$. write down the corresponding Lagrange's equations. (Rohilkhand 1996)

Sol.

Here $\dot{x} = \dfrac{dx}{dt} = \dfrac{dx}{ds}.\dfrac{ds}{dt} = x'\dot{s} \qquad [\, x' = \dfrac{dx}{ds} \text{ and } \dot{s} = \dfrac{ds}{dt} \,]$

And $\quad \dot{z} = \dfrac{dz}{dt} = \dfrac{dz}{ds}.\dfrac{ds}{dt} = z'\dot{s}$

Kinetic energy $T = \dfrac{1}{2}m\left(\dot{x}^2 + \dot{z}^2\right) = \dfrac{1}{2}m\left(x'^2 + z'^2\right)\dot{s}^2$

Potential energy $V = mgz$

Lagrange's equations are $\dfrac{d}{dt}\left(\dfrac{\partial L}{\partial \dot{s}}\right) - \dfrac{\partial L}{\partial s} = 0$

Here $\dfrac{\partial L}{\partial \dot{s}} = \dfrac{1}{2}\left(x'^2 + z'^2\right)2\dot{s} = \left(x'^2 + z'^2\right)\dot{s}$

And $\dfrac{\partial L}{\partial s} = \left[\dfrac{1}{2}m\dot{s}^2\left(2x'x'' + 2z'z''\right)\right] - mg\dfrac{\partial z}{\partial s}$

$\qquad = m\dot{s}^2\left(x'x'' + z'z''\right) - mgz'$

Hence $\dfrac{d}{dt}\left[m\left(x'^2 + z'^2\right)\dot{s}\right] - m\dot{s}^2\left(x'x'' + z'z''\right) + mgz' = 0$

This is the required Lagrange's equation.

3. A particle of mass m moves on a plane in the field of force given by (in polar) $\vec{F} = -kr\cos\theta\hat{r}$.

(a) Will the angular momentum of the particle about the origin be conserved? Justify your statement.

(b) Obtain the differential equation of the orbit of the particle.   (Agra-1995).

Solution.

Here $T = \dfrac{1}{2}\left(\dot{r}^2 + r^2\dot{\theta}^2\right)$

$\dfrac{\partial T}{\partial \theta} = 0; \ \dfrac{\partial T}{\partial \dot{\theta}} = mr^2\dot{\theta}; \ \dfrac{\partial T}{\partial r} = mr\dot{\theta}^2 \ \text{and} \ \dfrac{\partial T}{\partial \dot{r}} = m\dot{r}$

(a) $\dfrac{d}{dt}\left(\dfrac{\partial T}{\partial \dot{\theta}}\right) - \dfrac{\partial T}{\partial \theta} = G_\theta$

Since no transverse force is there, $G_\theta = 0$ so $\dfrac{d}{dt}\left(mr^2\dot{\theta}^2\right) = 0$. Hence the angular momentum about the origin is conserved.

(b) $\dfrac{d}{dt}\left(\dfrac{\partial T}{\partial \dot{r}}\right) - \dfrac{\partial T}{\partial r} = G_r$ or $m\ddot{r} - mr\dot{\theta}^2 = -kr\cos\theta$

which represents the equation of motion.

4. A cylinder of radius r and mass m rolls down on inclined plane making an angle Ø with horizontal. Set up the Lagrangian and find equation of motion.

Solution.

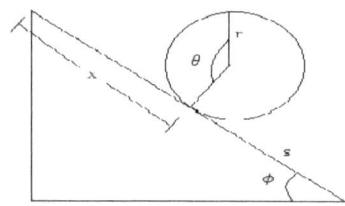

here $T = \dfrac{1}{2}m\dot{x}^2 + \dfrac{1}{2}I\omega^2$

$= \dfrac{1}{2}m\dot{x}^2 + \dfrac{1}{2}\dfrac{mr^2}{2} \qquad = \dfrac{3}{4}m\dot{x}^2$

As $I = \dfrac{mr^2}{2}$ and $\omega = \dot{\theta} = \dfrac{x}{r}$ and

$V = mg(s-x)\sin\theta + mgr\cos\theta$

So $L = \dfrac{3}{4}m\dot{x}^2 - mg(s-x)\sin\theta - mgr\cos\theta$

So equation of motion is

$$\frac{3}{2}m\dot{x}^2 - mg\sin\theta = 0$$

5. A bead slides on a smooth rod which is rotating about one end in a vertical plane with uniform angular velocity ω. Show that the equation of motion is $m\ddot{r} = mr\omega^2 - mg\sin\omega t$.

Solution.

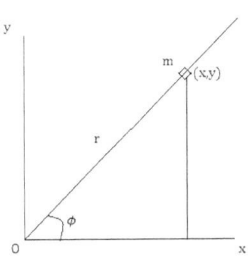

here $T = \frac{1}{2}m\left(\dot{r}^2 + r^2\dot{\phi}^2\right)$

and $V = mgy = mgr\sin\phi$

so $L = T - V = \frac{1}{2}m\left(\dot{r}^2 + r^2\dot{\phi}^2\right) - mgr\sin\phi$

so $\frac{\partial L}{\partial r} = mr\dot{\phi}^2 - mg\sin\phi$ and $\frac{\partial L}{\partial \dot{r}} = m\dot{r}$

so from $\qquad \frac{d}{dt}\left(\frac{\partial L}{\partial \dot{r}}\right) - \frac{\partial L}{\partial r} = 0$

we can get $\qquad m\ddot{r} - mr\dot{\phi}^2 + mg\sin\phi = 0$

or $\qquad m\ddot{r} - mr\omega^2 + mg\sin\omega t = 0 \qquad$ where $\dot{\phi} = \omega$ and $\phi = \omega t$.

6. Find out the Hamiltonian and Hamilton's equations for an arrangement of ideal spring-mass.

Solution.

Here $L = \frac{1}{2}m\dot{x}^2 - \frac{1}{2}kx^2 \qquad$ where k is the force constant of spring.

$P_x = \frac{\partial L}{\partial \dot{x}} = m\dot{x} \qquad$ or $\qquad \dot{x} = \frac{P_x}{m}$

And Hamiltonian $H = \sum_k p_k\dot{q}_k - L \qquad = p_x\dot{x} - \left[\frac{1}{2}m\dot{x}^2 - \frac{1}{2}kx^2\right]$

$$= \frac{1}{2}m\dot{x}^2 + \frac{1}{2}kx^2$$

Or $H = \dfrac{p_x^{\,2}}{2m} + \dfrac{1}{2}kx^2$

So Hamiltonian Equations are

$\dot{x} = \dfrac{\partial H}{\partial p_x} = \dfrac{p_x}{m}$ and $\dot{p}_x = -\dfrac{\partial H}{\partial x} = -kx$

So equation of motion

$m\ddot{x} = -kx$ or $\ddot{x} + \dfrac{k}{m}x = 0$

Which is the required Hamilton's equation for the given arrangement.

7. Find equation of Hamilton's for a particle falling freely under gravity.

$\hfill$ (WBUT-2005)

Solution.

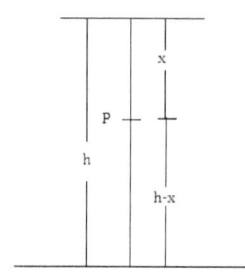

Here Potential Energy $V = mg(h - x)$

Kinetic Energy $T = \dfrac{1}{2}m\dot{x}^2$

So Lagrangian   L=T-V

$\hspace{3cm} = \dfrac{1}{2}m\dot{x}^2 - mg(h - x)$

So $p_x = \dfrac{\partial L}{\partial \dot{x}} = m\dot{x}$ or $\dot{x} = \dfrac{p_x}{m}$

And Hamiltonian $H = \sum p_x \dot{x} - L = \dfrac{p_x^{\,2}}{m} - \left[\dfrac{1}{2}m\dfrac{p_x^{\,2}}{m^2} - mg(h - x) \right]$

$\hspace{3cm} = \dfrac{p_x^{\,2}}{2m} + mg(h - x)$ = total energy of the system

So Hamilton's equations are

$\dot{p}_x = -\dfrac{\partial H}{\partial x} = mg$ and $\dot{x} = \dfrac{\partial H}{\partial p_x} = \dfrac{p_x}{m}$

Or $p_x = m\dot{x}$ or $\dot{p}_x = m\ddot{x}$

Or $mg = m\ddot{x}$ or $\ddot{x} = g$ or $\ddot{x} - g = 0$

Which is the required equation of motion.

8. An Eigen function of an operator $\dfrac{d^2}{dx^2}$ is $\psi = e^{ax}$. Calculate the Eigen values.

Solution.

Let $F = \dfrac{d^2}{dx^2}$ so $F\psi = \dfrac{d^2}{dx^2}\left(e^{ax}\right)$ $= \dfrac{d}{dx}\left[\dfrac{d}{dx}\left(e^{ax}\right)\right] = a^2 e^{ax}$

But $e^{ax} = \psi$ so $F\psi = a^2\psi$ or $F = a^2$

9. A particle limited to x axis with wave function $\psi = ax$ between x =0 to 1 and $\psi = 0$ in other regions. Find the probability that the particle can found between x =0.45 and x =0.55.

Solution.

We know that the Probability is given by $P = \int\limits_{x_1}^{x_2} |\psi|^2\, dx$

So $= \int\limits_{0.45}^{0.55} a^2 x^2 dx = a^2 \dfrac{x^3}{3}\bigg|_{0.45}^{0.55}$

Hence $P - 0.0251a^2$

10. The wave function of a certain particle is $\psi = A\cos^2 x$ for $-\dfrac{\pi}{2} < x < \dfrac{\pi}{2}$. (a) find A and (b) find probability for particle to be in x =0 and $x = \dfrac{\pi}{4}$.

Solution.

As $\psi = A\cos^2 x$

(a) $\int\limits_{-\frac{\pi}{2}}^{\frac{\pi}{2}} |\psi|^2\, dx = 1 \Rightarrow 2A^2 \int\limits_{0}^{\frac{\pi}{2}} \cos^4 x dx = 1$ or $2A^2 \dfrac{3\pi}{16} = 1 \Rightarrow \dfrac{3\pi}{8} A^2 = 1$

So $A = \sqrt{\dfrac{8}{3\pi}}$

(b) $P = \int\limits_{0}^{\frac{\pi}{4}} |\psi|^2\, dx = A^2 \int\limits_{0}^{\frac{\pi}{4}} \cos^4 x dx$

As $A = \sqrt{\dfrac{8}{3\pi}}$ so $P = \dfrac{8}{3\pi} \int\limits_{0}^{\frac{\pi}{4}} \cos^4 x dx = 0.462$

11. Calculate the energy of an electron moving in one dimension in an infinitely high potential box of width 1Å.

Solution.

Energy is given by $E = \dfrac{n^2 h^2}{8mL^2}$ where n=1,2,3,....

Least energy is for n =1

So $E = \dfrac{h^2}{8mL^2} = \dfrac{\left(6.63\times10^{-34}\right)^2}{8\times9.11\times10^{-31}\times\left(10^{-10}\right)^2}$joules

$= 9.1\times10^{-19}$ joules $= = \dfrac{9.1\times10^{-19}}{1.6\times10^{-19}}$ eV $= 5.68$ eV

12. Compute the lowest energy of a neutron confined in a nucleus. Given size of nucleus = 10^{-14} m across.

Solution.

As energy is given by $E_{n_x,n_y,n_z} = \dfrac{\pi^2 \hbar^2}{2m}\left[\dfrac{n_x^2}{a^2}+\dfrac{n_y^2}{b^2}+\dfrac{n_z^2}{c^2}\right]$

So lowest energy is when n_z, n_y and n_z =1

i.e. $E_{1,1,1} = \dfrac{\pi^2 \hbar^2}{2m}\left(\dfrac{3}{a^2}\right) = \dfrac{3\times\left(6.625\times10^{-34}\right)^2}{8\times1.6\times10^{-27}\times10^{-28}}$

$$= 10.29 \times 10^{-23} J$$
$$= 6.43 MeV$$

13. A particle is trapped in a box of width 25Å. Find the probability within an interval of 5Å at the centre of the box with the lowest energy value.

Solution:

Wave function for one dimensional box of side L is given by

$$\psi = \sqrt{\frac{2}{L}} \sin\frac{n\pi}{L}x$$

For lowest energy n=1, at the centre x = L/2
So probability

$$|\psi_x|^2 = \left[\sqrt{\frac{2}{L}} \sin\frac{\pi\left(\frac{L}{2}\right)}{L}\right]^2 = 2/L$$

For Δx region, probability

$$P = |\psi|^2 \Delta x = \frac{2}{L}\Delta x = \frac{2 \times 5 \times 10^{-10}}{25 \times 10^{-10}}$$

$$= 0.4$$

14. Evaluate the expectation value of kinetic energy of a particle in one dimensional rigid box in nth quantum state.

Solution:

Expectation value for kinetic energy is given by

$$\langle E \rangle = \int_{-\infty}^{\infty} \psi^* E_k \psi\, dx \quad = \int_{-\infty}^{\infty} \psi^* \frac{p^2}{2m} \psi\, dx$$

$$= \frac{1}{2m} \int_{-\infty}^{\infty} \psi^* (-i\hbar)^2 \frac{\partial^2}{\partial x^2}\psi\, dx$$

$$= \frac{\hbar^2}{2m} \int_{-\infty}^{\infty} \psi^* \frac{\partial^2}{\partial x^2}\psi\, dx$$

$$= \frac{2\hbar^2}{2ma} \int_0^a \sin\frac{n\pi x}{a}\left[-\frac{n^2\pi^2}{a^2}\right]\sin\frac{n\pi x}{a}dx$$

$$= \frac{n^2\pi^2\hbar^2}{2ma^3}.a$$

$$= \frac{n^2\pi^2\hbar^2}{2ma^2} = \frac{n^2h^2}{8ma^2}$$

15. Assuming that proton is inside the nucleus. What amount of energy it will release if it transits from first excited state to ground state? The size of nucleus is 1.0 x 10⁻¹⁴ m.

Solution:

As $E_n = \dfrac{n^2 h^2}{8ma^2}$

Hence as per the given question

$$\Delta E = \frac{h^2}{8ma^2}\left(2^2 - 1^2\right)$$

$$= \frac{3\times\left(6.625\times10^{-34}\right)^2}{8\times1.6\times10^{-27}\times\left(10^{-14}\right)^2}$$

$$= \frac{3h^2}{8ma^2} = 10.3\times10^{-13}\,J$$

$$= 6.4\,MeV$$

16. In a long chain molecule of 5 Å, electron may be treated to be free to move along the length. Calculate zero point energy, the energy gap between first two energy state of the electrons, and also the wavelength of absorption line arising from this transition.
(MREC1BE2002)

Solution:

Zero point energy is $E_1 = \dfrac{h^2}{8mL^2} = \dfrac{\left(6.62\times10^{-34}\right)^2}{8\times9.1\times10^{-31}\times\left(5\times10^{-10}\right)^2}$

$$=2.4\text{x}10^{-19}\,J$$

$$=1.5\text{ eV}$$

Energy gap $\Delta E = 3.\dfrac{h^2}{8mL^2} = 3\times1.5 = 4.5eV$

Wavelength $\lambda = \dfrac{hc}{\Delta E} = \dfrac{hc}{4.5eV} = \dfrac{6.62\times10^{-34}\times3\times10^3}{4.5\times1.6\times10^{-19}}$

$$=2.76\text{x}10^{-7}$$

$$=2760\,Å$$

Module – 6

Statistical Mechanics

Concept of energy levels and energy states, Microstates and Macrostates and thermodynamic probability, Equilibrium Macrostates, MB, FD, BE statistics (no derivation necessary), Fermions, Bosons (definition in terms of spin, examples), Physical Significance and applications, classical limits of quantum statistics, Fermi distribution at zero and non-zero temperatures, calculation of Fermi level in metals and also total energy at absolute zero of temperature and total number of particles, Bose-Einstein statistics – Planck's law of Black body radiation.

INTRODUCTION

When we deal with problems related to macroscopic systems i.e. which are in ordinary dimensions, then statistical physics work to solve them as the basic fundamental parameter is the temperature of the system. As temperature of a body is related to kinetic energy of the random motion of micro-particles around the centre of mass of the system, the concept of temperature is purely statistical and it is not valid for a single particle or some particles. This is based on Maxwell theory of kinetic energy in 1860. Statistical approach can be applied only to those systems which contain a very large number of identical particles which can be distinguished by means of some property. Statistical mechanics provides probability laws for distribution of these identical microscopic particles of a macroscopic system under thermal equilibrium. It does not care about detailed study of position and velocity of each particle and calculating equilibrium properties of the system. In statistics mechanics, it is assumed that properties and laws of motion for each particle are known. When these laws are governed by classical mechanics, it is called classical statistics and when these laws are governed by quantum mechanics, it is known as quantum statistics. Hence it can be divided in to following

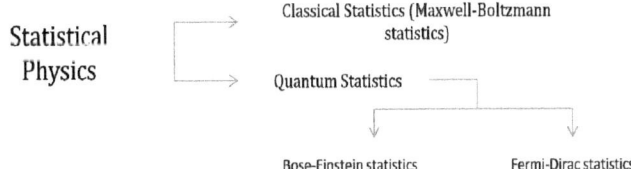

Maxwell, while studying dynamical theory of gases or classical particles gave classical statistics in 1860 which was further enhanced by Boltzmann and known as Maxwell-Boltzmann statistics. Quantum particles are of two types (i) with integral spin (i.e. $n\hbar$) known as bosons and (ii) with half-integral spin $\left[\left(n+\dfrac{1}{2}\right)\hbar\right]$ known as Fermions. Related statistical studies are Bose-Einstein and Fermi-Dirac statistics.

Classical statistics differs from quantum statistics in its application of principle of indistinguishability of identical particles. Classical statistics explains successfully many physical phenomenons (e.g. temperature, pressure etc) of any system, but it fails to explain black body radiation or specific heat at low temperature of a system. These phenomenons are explained well using quantum statistics. At higher temperature and low pressure Maxwell-Boltzmann, Bose-Einstein and Fermi-Dirac give the same result.

Basic Concepts

1. States

According to classical mechanics, a 'state' of any system is defined by the values of corresponding coordinates and momenta of all particles. Every separate particle has degree of freedom which corresponds to energy $\left(\dfrac{1}{2}kT\right)$, where k is Boltzmann constant. In classical Mechanics, all degree of freedom are equally justified with respect to energy.

In quantum statistics, a large number of particles obeying quantum laws are studied. Assuming that the exact description is according to the corresponding probability i.e. wave function of coordinates and time. Quasi-classical approximation deals with arbitrary quantum mechanical systems having degrees of freedom. Each state of particle which is associates with definite energy is called energy level of that. If there are many energy states which corresponds to one energy level then that energy level is said to be degenerate, and number of states associated with it is called its degeneracy.

2. Phase Space

According to classical mechanics, any particle in three dimensional space can be located using Cartesian coordinates x,y,z, whereas its velocity components v_x, v_y, v_z determines its state of motion. State of motion can also be represented using components of momentum p_x, p_y and p_z. Phase space is defines as this six dimensional space, which is required to describe particle's position as well as its state of motion. It has six coordinates x, y, z, p_x, p_y, p_z. Any point if phase space represents its position and its motion. For a system of N particles, 3N positions coordinates and 3N momentum coordinates are used to determine its mechanical state. This 6N dimensional space is called phase space or Γ space of the system. Any point in Γ space represents a state of the entire system. For mono-atomic gas, this space is called as molecular phase space or μ - space. So any volume element in μ - space is given by $d\tau = dxdydzdp_x dp_y dp_z$. such a cell of minimum volume is called a unit cell in μ-space.

Hence $d\tau = (\Delta x \Delta p_x)_{\min} (\Delta y \Delta p_y)_{\min} (\Delta z \Delta p_z)_{\min}$

And according to Heisenberg's uncertainty principle

$\Delta x \Delta p_x \approx \Delta y \Delta p_y \approx \Delta z \Delta p_z \approx h$ (Planck's constant)

So $d\tau = h.h.h = h^3$

Hence phase space is actually a cell of minimum volume h^3.

3. Microstate

Microstate of any system is the complete specification of set of six coordinates (x, y, z, p_x, p_y, p_z) of each particle of a system. The fundamental hypothesis of statistical mechanics is that all microstates are equally probable.

4. Macrostate

It is defines as number of phase points in each cell of the phase space. For this only those quantities, which can be determined by macroscopic measurements, can be specified like pressure, temperature etc.

e.g. let us assume there are four distinguishable particles p,q,r,s which are to be distributed in to two exactly similar boxes. As both boxes exactly alike, hence particles will have same priority probability of getting into the boxes. Hence distribution can be as below-

Macrostate	Possibility of arrangements		Microstates
	Box 1	Box 2	
0,4	0	pqrs	1
1,3	p	prs	4
	q	rsp	
	r	spq	
	s	pqr	
2,2	pq	rs	6
	pr	qs	
	ps	qr	
	qr	ps	
	qs	pr	
	rs	pq	
3,1	pqr	s	4
	prs	q	
	qrs	p	
	pqs	r	
4,0	pqrs	0	1

5. Thermodynamic Probability and Entropy

Thermodynamic probability is the number of possible microstates corresponding to any given macrostate. It is very large in comparison to mathematical probability. The entropy and thermodynamic probability is related as

Entropy $S = K_B \ln W$ where K_B = Boltzmann constant and W = thermodynamic probability.

6. Ensembles

In 1902, French scientist Gibbs proposed a different approach to solve statistical physics problems. In statistical mechanics, in place of large number of phase variables, a huge set of those systems is considered. That group of systems or assembly of systems is called an Ensemble. Such an arrangement of Ensemble is macroscopically identical but microscopically they are different. Macroscopic states such as pressure, volume, temperature, entropy etc. are lesser then dynamical microscopic states. According to the interaction of systems, the Ensembles are classified in to three categories –
(a) Canonical Ensemble – when there is exchange of energy but not matter between systems, then ensemble is called canonical ensemble.

(b) Grand Canonical Ensemble – when both energy and matter are exchanged in between systems, then it is grand canonical ensemble.

(c) Micro Canonical Ensemble – when neither energy nor matter is exchanged, then it is micro canonical ensemble.

7. Fermions

Indistinguishable particles with half-integral spin obeying Fermi-Dirac statistics are called Fermions. E.g. electrons ($m_s = \pm\dfrac{1}{2}$), positrons ($m_s = \pm\dfrac{1}{2}$), neutrons ($m_s = \pm\dfrac{1}{2}$), protons ($m_s = \pm\dfrac{1}{2}$), μ-mesons ($m_s = \pm\dfrac{1}{2}$),. Experimentally it has been observed that atoms with odd numbers of nucleons obey Fermi-Dirac statistics. Hence $_1H^3, _2He^3, _3Li^7, _6C^{13}$ are also can be considered as Fermions.

8. Bosons

Indistinguishable, identical particles with integral spin $\left(m_s = 0, \hbar, 2\hbar....\right)$ obeying Bose-Einstein statistics are called Bosons. Further if atom as even number of nucleons, they obey Bose-Einstein statistics. E.g. photons (1), phonons (1), π-mesons (0), α-particle (0), K-mesons, η-mesons and $_1H^2, _2He^4, _6C^{12}, _8O^{16}$.

9. Equilibrium Macrostate

When any macrostate of a system does not change with respect to time then it is called equilibrium macrostate. At equilibrium, macrostate is time-independent except for random fluctuations. It is also independent of its historical developments. As macrostate describes the macroscopic parameters of a system, it is independent of microscopic changes.

Physical significance of MB, BE and FD statistics

In statistical mechanics, Maxwell-Boltzmann (MB) statistics describes the statistical distribution of particles over various energy states in thermal equilibrium, with very high temperature and very low density to make quantum effects negligible. But when quantum effects are effective then Fermi-Dirac (FD) and Bose-Einstein (BE) statistics can be applied. In general quantum effects appears to work if concentration of particles $\left(\dfrac{N}{V}\right) >> n_q$ where n_q is the quantum concentration which is when the inter particle distance is equal to thermal de-Broglie wavelength i.e. when the wave function of particles are not overlapping but only touching each other.

Maxwell-Boltzmann (MB) statistics

Basic postulates-

(i)   Particles are weakly interacting, identical and distinguishable.

(ii)   Particles do not have any kind of spin.

(iii) Particles do not obey Pauli's exclusion principle so each state can accommodate any number of particles.

4

Maxwell-Boltzmann statistics gives statistical behavior of distinguishable and identical spinless particles of any ideal gas which do not obey Pauli's exclusion principle. The particles obeying MB statistics are called Boltzons.

If N are the total number of distinguishable particles and g_i is the probability of locating a particle in a certain energy state E_i then total probability

$$W = \underline{|N|} \prod_i \frac{(g_i)^{N_i}}{\underline{|N|}}$$ ----------------(i)

Where $N_i = g_i e^{-(\alpha + \beta E_i)}$ ----------------(ii)

Equation (ii) is Maxwell Boltzmann distribution law,

Or $$N_i = \left(\frac{N}{Z}\right) g_i e^{-\beta E_i}$$ ----------------(iii)

Where $Z = \sum_i g_i e^{-\beta E_i}$ is called the partition function. The values of α and β depends upon the physical properties of the system.

Physical significance and applications-

Maxwell distribution law can be applied to calculate average speed, root mean square speed, most probable speed of molecules of ideal gas, total internal energy, specific heat at constant volume of an ideal gas etc.

Limitations/Drawbacks of MB statistics

- For a given density, the temperature of an ideal gas becomes sufficiently low, where MB statistics cannot be applied.
- Here particles are assumed to be distinguishable, but in actual practice many particles like electrons are indistinguishable.
- Here particles do not obey Pauli's exclusion principle, but in practice many particles obey that and restrict number of particles in a particular state.

All the above drawbacks are overcome by using quantum statistics.

Bose-Einstein (BE) statistics

Basic postulates-

(i)   Here the particles of the system in consideration are identical and indistinguishable.
(ii)   Here particles do not obey Pauli's exclusion principle hence any energy state can accommodate any number of particles.
(iii)   Bose-Einstein statistics can be applied only to particles having integral spin angular momentum i.e. $m_s \hbar = 0, \hbar, 2\hbar ....$
(iv)   These particles have symmetric wave functions.

5

Bose-Einstein statistics is used for indistinguishable, identical particles with integral spin. These do not obey Pauli's exclusion principle. These are known as Bosons.

Let us consider total number of ways, by which N_i identical, indistinguishable particles can be distributed among g_i quantum states of energy E_i, such that any number of particles can occupy any state; given by

$$W = \prod_{i=1}^{n} W_i = \prod_{i=1}^{n} \frac{\lfloor N_i + g_i - 1}{\lfloor N_i \lfloor g_i - 1}$$
----------------(iv)

As N_i and g_i are very large compared to unity so

$$W = \prod_{i=1}^{n} \frac{\lfloor N_i + g_i}{\lfloor N_i \lfloor g_i}$$
----------------(v)

And
$$N_i = \frac{g_i}{e^{\alpha + \beta E_i} - 1}$$
----------------(vi)

Equation (vi) represents the Bose-Einstein distribution law. It gives the most probable distribution of bosons among the various energy levels.

In thermal equilibrium at temperature T, we can write

$\alpha = -\dfrac{\mu}{kT}$ and $\beta = \dfrac{1}{kT}$ where μ is chemical potential of the system.

So
$$N_i = \frac{g_i}{e^{(E_i - \mu)/kT} - 1}$$
----------------(vii)

Hence Bose-Einstein distribution function can be written as

$$f(E) = \frac{N_i}{g_i} = \frac{1}{e^{(E_i - \mu)/kT} - 1}$$
----------------(viii)

And if $E_i \gg kT$, BE statistics reduces to MB statistics.
Here k = Boltzmann constant = 8.62×10^{-5} eV/K = 1.3805×10^{-23} J.K^{-1}

According to BE statistics, boson particles for different energy levels can be shown like

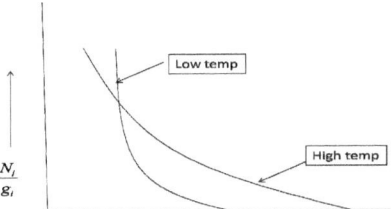

Application of BE statistics (Derivation of Planck's law of Black body radiation)-

As proposed by Einstein, black body radiations are composed of discrete energy packets called photons. The radiations inside an enclosure at constant temperature can be considered equivalent to photon gas, where walls of blackbody can absorb or emit photons of different energies. These photons are indistinguishable with many photons having same energy. These photons can interact with only walls of blackbody. Hence total numbers of photons are constant.

So $\sum_i N_i$ = constant or $\sum_i dN_i = 0$ is not valid for photon gas although the total energy of the photons remain constant. So in this case $\alpha=0$ so from eq. (vi)

$$N_i = \frac{g_i}{e^{\beta E_i} - 1} = \frac{g_i}{e^{E_i/kT} - 1} \qquad \text{----------------(ix)}$$

Energy of each photon was given by Einstein as E=hv

$$N_i = \frac{g_i}{e^{hv/kT} - 1} \qquad \text{----------------(x)}$$

The number of photons in the frequency range v and v+dv is obtained by replacing g_i by $g(v)dv$ and N_i by $N(v)dv$. Hence

$$N(v)dv = \frac{g(v)dv}{e^{hv/kT} - 1} \qquad \text{----------------(xi)}$$

Where $g(v)dv$ is the number of quantum states in the frequency range v and v+dv. Now the probability g(p) that a molecule has a momentum between p and p+dp is equal to the number of cells in phase space. If each cell has volume h^3 then

$$g(p)dp = \frac{\iiint dxdydz \iiint dp_x dp_y dp_z}{h^3}$$

here $\iiint dxdydz = V$ volume of gas in ordinary position space

and $\iiint dp_x dp_y dp_z = 4\pi p^2 dp$ volume of spherical shell of radius p with thickness dp.

7

So $\quad g(p)dp = \dfrac{4\pi p^2 V dp}{h^3}$

Now every system contains two types of identical particles; one with clockwise spin and other with anti-clockwise spin. So

$$g(p)dp = 2 \times \frac{4\pi p^2 V dp}{h^3} = \frac{8\pi p^2 V dp}{h^3} \quad\text{------------(xii)}$$

Now E = hv and p = hv/c, so putting these values in above eq. (xii)

$$g(v)dv = \frac{8\pi V \left(\dfrac{hv}{c}\right)^2 \left(\dfrac{h}{c}\right) dv}{h^3}$$

$$= \frac{8\pi V v^2 dv}{c^3} \quad\text{------------(xiii)}$$

Hence from eq. (xi)

$$N(v)dv = \frac{8\pi V v^2 dv}{c^3} \cdot \frac{1}{e^{hv/kT} - 1} \quad\text{------------(xiv)}$$

The energy density i.e. energy per unit volume within v and v+dv is given by

$$U_v dv = \frac{hv}{V} N(v)dv \quad\text{------------(xv)}$$

Or $\quad \boxed{U_v dv = \dfrac{8\pi hv^3}{c^3} \dfrac{dv}{e^{hv/kT} - 1}} \quad\text{------------(xvi)}$

This is Planck's law of radiation in terms of frequency v.

Fermi – Dirac (FD) statistics-

Basic postulates-

(i) Here the particles are identical and indistinguishable.
(ii) On these particles, Pauli's exclusion principle is applicable so each quantum state can hold either no particle or only one particle.
(iii) These particles have half integral spin angular momentum in terms of h.
(iv) As no particle has same quantum state, these particles have anti-symmetric wave function.

Fermi-Dirac statistics was developed by E. Fermi and P. Dirac to find the energy distribution among indistinguishable identical particles having half integral spin.

Let us consider the total number of possible ways in which N_i number of identical, indistinguishable particles can occupy g_i sublevels. So probability of the entire distribution is given by

$$W = \prod_i \frac{g_i}{\left| N_i \right| g_i - N_i}$$ ----------------(i)

And $$N_i = \frac{g_i}{e^{\alpha+\beta E_i}+1}$$ ----------------(ii)

Eq. (ii) is Fermi-Dirac distribution law. α and β can be taken for fermions in statistical equilibrium at T, $\alpha = -\dfrac{E_f}{kT}$ and $\beta = \dfrac{1}{kT}$ where E_f = Fermi energy of the system

So $$N_i = \frac{g_i}{e^{(E_i-E_f)/kT}+1}$$ ----------------(iii)

And FD distribution function is

$$\boxed{f(E) = \frac{N_i}{g_i} = \frac{1}{1+e^{(E_i-E_f)/kT}}}$$ ----------------(iv)

Fermi Distribution at zero and non-zero temperature

By plotting graph between $f(E)$ with E_i at different temperature. FD distribution function which gives the average occupation of energy level is

$$f(E) = \frac{1}{1+e^{(E_i-E_f)/kT}}$$ ----------------(v)

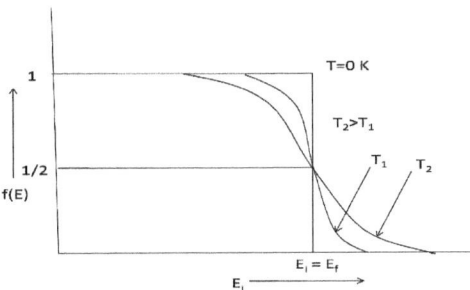

Case-1 – when T = 0

f(E) has two values possibly

if $E_i < E_f$, $f(E) = \dfrac{1}{1+e^{-\infty}} = 1$

and if $E_i > E_f$, $f(E) = \dfrac{1}{1+e^{\infty}} = 0$

hence at absolute zero (T=0 K) all possible quantum states of energy less then E_f are occupied and all with more energy are empty. So Fermi energy is defined as the energy of the highest occupied level at absolute zero or it is the maximum energy that can be occupied by a fermion at 0 K.

Case-2 – when T>0 and $E_i = E_f$

$$f(E) = \frac{1}{1+e^{(E_i - E_f)/kT}} = \frac{1}{2}$$

So for Fermi energy level the probability of occupation is ½ at any temperature greater than 0 K.

Fermi Temperature

It is the ratio of Fermi Energy (E_f) at absolute zero to Boltzmann constant (k) and its corresponding absolute temperature i.e. $\theta_f = \dfrac{E_f}{kT}$.

Fermi-Dirac Law for free electrons in Metals

In metals there is overlapping of valence band and conduction band, due to which, a large number of free electrons are present in metals making it a good conductor. These free electrons behave like a gas, called electron gas. Considering the distribution of energy as continuous among N free electrons in electron gas then for energy range E and E+dE, total number of electrons can be written as

$$n(E)dE = \frac{g(E)dE}{e^{(E-E_f)/kT}+1} \qquad \text{-----------------(i)}$$

Where g(E)dE is the number of energy states within energy states range E and E+dE. Then in terms of momentum

$$g(p)dp = \frac{8\pi V}{h^3} p^2 dp = \frac{8\pi V}{h^3} p(pdp) \qquad \text{---------------(ii)}$$

Now $E = \frac{1}{2}mv^2 = \frac{p^2}{2m}$ or $p^2 = 2mE$

Or $2pdp = 2mdE$ or $pdp = mdE$ \qquad \text{---------------(iii)}

Using (iii) in (ii)

$$g(p)dp = \frac{8\pi V}{h^3}(2mE)^{\frac{1}{2}} mdE \qquad \text{---------------(iv)}$$

Now from (iv) and (i)

$$\boxed{n(E)dE = \frac{8\sqrt{2}}{h^3} \frac{m^{\frac{3}{2}} E^{\frac{1}{2}} dE}{e^{(E-E_f)/kT}+1}} \qquad \text{---------------(v)}$$

Eq. (v) is the Fermi-Dirac law of energy distribution for free electrons in metals.

Total number of particles and total energy at absolute zero of temperature

We know that at absolute zero of temperature, the total number of electrons (N) is equal to the total number of energy states occupied by electron from 0 to E_f since each state can have only one electron.

So $N - \int_0^{E_f} n(E)dE = \int_0^{E_f} g(E)dE$ \qquad \text{[as f(E) =1]}

$$= 4\pi V \left(\frac{2m}{h^2}\right)^{\frac{3}{2}} \int_0^{E_f} E^{\frac{1}{2}} dE$$

So $\boxed{N = \frac{8\pi V}{3}\left(\frac{2mE_f}{h^2}\right)^{\frac{3}{2}}}$ \qquad \text{---------------(vi)}

This is the total number of electrons in a metal at absolute zero of temperature.

Eq. (vi) can also be written as

$$E_f = \frac{h^2}{2m}\left(\frac{3N}{8\pi V}\right)^{\frac{2}{3}}$$

If $n = \dfrac{N}{V}$ = number of free electrons per unit volume i.e. free electron density.

Then $\boxed{E_f = \dfrac{h^2}{2m}\left(\dfrac{3n}{8\pi}\right)^{\frac{2}{3}}}$ ----------------(vii)

Eq.(vii) represents the Fermi energy at absolute zero of temperature, according to this equation Fermi energy only depends upon free electrons per unit volume i.e. (N/V) and is independent of size or volume of the conductor.

Average energy of free electrons at absolute zero in a metal

Average energy of free electrons in a metal at absolute zero

$$\overline{E} = \frac{1}{N}\int_0^\infty En(E)dE$$

$$= \frac{1}{N}\int_0^{E_f} Ef(E)g(E)dE + \frac{1}{N}\int_{E_f}^\infty Ef(E)g(E)dE$$

But at T=0 K , f(E) =1 if E<E_f and f(E) =0 if E>E_f

So $\quad \overline{E} = \dfrac{1}{N}\int_0^{E_f} Eg(E)dE$

And $\quad g(E)dE = 4\pi V\left(\dfrac{2m}{h^2}\right)^{\frac{3}{2}} E^{\frac{1}{2}}dE$

So $\quad \overline{E} = \dfrac{1}{N}4\pi V\left(\dfrac{2m}{h^2}\right)^{\frac{3}{2}}\int_0^{E_f} E^{\frac{3}{2}}dE$

$$= \left[\frac{4\pi V}{N}\left(\frac{2m}{h^2}\right)^{\frac{3}{2}}\frac{2}{5}E_f^{\frac{3}{2}}\right]E_f$$ ----------------(i)

But at 0 K

$$E_f = \frac{h^2}{2m}\left(\frac{3N}{8\pi V}\right)^{\frac{2}{3}} \quad \text{or} \quad E_f^{\frac{3}{2}} = \left(\frac{h^2}{2m}\right)^{\frac{3}{2}}\left(\frac{3N}{8\pi V}\right) \text{---------------(ii)}$$

Using (ii) in (i)

$$\boxed{\overline{E} = \frac{3}{5}E_f}$$ ----------------(iii)

So average electron energy is $3/5^{th}$ of the Fermi energy at absolute zero.

Fermi Velocity and Average Velocity of free electrons at zero Kelvin

As we know that Energy at absolute zero = kinetic energy

i.e. $$E_f = \frac{h^2}{2m}\left(\frac{3n}{8\pi}\right)^{\frac{2}{3}} = \frac{1}{2}mv_f^2$$ where v_f is Fermi velocity.

So $$\boxed{v_f = \frac{h}{m}\left(\frac{3n}{8\pi}\right)^{\frac{1}{3}}}$$ ----------------(i)

If $\overline{v}$ is the average speed of an electron at 0 K, then

$$\overline{v} = \frac{1}{N}\int_0^{v_f} vn(v)dv$$ ----------------(ii)

Where $n(v)dv$ is the number of particles within the velocity range v and v+dv.

Now $F = \frac{1}{2}mv^2$ or dE=mvdv

Again at 0 K, f(E) = 1 if E<E_f so n(E)dE = g(E)dE

$$= \frac{8\pi V}{h^3}(\sqrt{2mE})mdE$$

So $$n(v)dv = 4\pi V\left(\frac{2m}{h^2}\right)^{\frac{3}{2}}\left(\frac{1}{2}mv^2\right)^{\frac{1}{2}}(mvdv)$$

$$= 8\pi V\left(\frac{m}{h}\right)^3 v^2 dv$$ ----------------(iii)

Now using (iii) in (ii)

$$\bar{v} = \frac{1}{N} \int_0^{v_f} 8\pi V \left(\frac{m}{h}\right)^3 v^3 dv$$

$$= \frac{8\pi V}{N} \left(\frac{m}{h}\right)^3 \left(\frac{v_f^4}{4}\right)$$

$$\frac{3}{4}\left(\frac{8\pi}{3n}\right)\left(\frac{m}{h}\right)^3 v_f^4$$

But $v_f = \frac{h}{m}\left(\frac{3n}{8\pi}\right)^{\frac{1}{3}}$ so from (ii)

$$\bar{v} = \frac{3}{4}\left(\frac{8\pi}{3n}\right)\left(\frac{m}{h}\right)^3 \left(\frac{h}{m}\right)^3 \left(\frac{3n}{8\pi}\right) v_f$$

Or $\boxed{\bar{v} = \frac{3}{4}v_f}$ -----------------(iv)

Hence average speed is equal to $3/4^{th}$ of the Fermi velocity at absolute zero.

SOLVED EXAMPLES

1. Distribute three particles in two different states according to (i) MB (ii) BE statistics.

Solution:

(i)   As per MB statistics particles are distinguishable, so total microstates

$$W = \frac{\lfloor Ng_i^{n_i}}{\lfloor n_i} = \lfloor 3 \frac{2^3}{\lfloor 3} \text{ as } n_i = 3, N = 3, g_i = 2$$

$$= (3 \times 2 \times 1) \times \frac{2 \times 2 \times 2}{3 \times 2 \times 1} = 8 \text{ microstates}$$

Let P,Q,R are three particles then energy states 1,2 will have as below

Energy state 1	Energy state 2
PQR	0
0	PQR
PQ	R
R	PQ
P	QR

Q	PR
QR	P
PR	Q

So total microstates =8

(ii)   BE statistics
Here particles are indistinguishable, so

$$W = \prod_{i=1}^{n} \frac{\lfloor N_i + g_i - 1}{\lfloor N_i \lfloor g_i - 1}$$

$$= \frac{\lfloor 3+2-1}{\lfloor 3 \lfloor 2-1} = \frac{\lfloor 4}{\lfloor 3 \lfloor 1} = 4 \text{ microstates}$$

As three particles are indistinguishable. Let us consider them as P. so

Energy state 1	Energy state 2
PPP	0
0	PPP
PP	P
P	PP

So total microstates = 4

(iii)   FD statistics

Here particles are indistinguishable and obey Pauli's exclusion principle

$$W = \frac{\lfloor g_i}{\lfloor n_i \lfloor g_i - n_i} = \frac{\lfloor 2}{\lfloor 3 \lfloor 2-3} = -ve \text{ which is not possible. So FD statistics can not be}$$

applied on these particles.

2.   Three distinguishable particles each of which can be accommodated in energy states E, 2E, 3E. 4E with total energy 6E. Find all the possible number of distributions. Also find total microstates in each case.   (WBUT-2007).

Solution:

Suppose three particles are P,Q,R and distinguishable.

Possible Arrangements	E	2E	3E	4E	Total Energy	Total Microstates
2,0,01	PQ	0	0	R	6E	3
	PR	0	0	Q	6E	
	QR	0	0	P	6E	
0,3,0,0	0	PQR	0	0	6E	1

15

1,1,1,0	P	Q	R	0	6E	6
	Q	P	R	0	6E	
	R	P	Q	0	6E	
	Q	R	P	0	6E	
	R	Q	P	0	6E	
	P	R	Q	0	6E	

Hence total arrangements = Macrostates = 3

And total microstates = 10

3. A system with non-degenerate single particle state with 0,1,2,3 energy units. Three particles are to be distributed in three states so that the total energy of the system is 3 units. Find the number of microstates if particles obey (i) MB statistics (ii) FD statistics.   (WBUT 2008).

Solution:

Let the particles be P,Q,R which are distinguishable. So for (i) MB statistics

Macrostates	0E	1E	2E	3E	Total Energy	Total Microstates
0,3,0,0	0	PQR	0	0	3E	1
2,0,0,1	PQ	0	0	R	3E	3
	PR	0	0	Q	3E	
	QR	0	0	P	3E	
1,1,1,0	P	Q	R	0	3E	6
	Q	P	R	0	3E	
	R	P	Q	0	3E	
	Q	R	P	0	3E	
	R	Q	P	0	3E	
	P	R	Q	0	3E	

So total microstates = 10
And for (ii) BE statistics- here particles are indistinguishable. suppose there are denoted by P

Macrostates	0E	1E	2E	3E	Total Energy	Total Microstates
0,3,0,0	0	PPP	0	0	3E	1
2,0,0,1	PP	0	0	P	3E	1
1,1,1,0	P	P	P	0	3E	1

So total microstates = 3

(iv)   FD statistics – here particles obey Pauli's exclusion principle and they are indistinguishable. So

Macrostates	0E	1E	2E	3E	Total Energy	Total Microstates
1,1,1,0	P	P	P	0	3E	1

So total microstates = 1

4. Show that at T=0, the average energy $\overline{E}$ of an electron in a metal is given by $\overline{E} = \dfrac{3}{5}E_f(0)$ where $E_f(0)$ is Fermi energy at absolute zero.

Solution:

As electron is a fermion. Total energy at absolute zero is given by

$$U(0) = \int_0^{E_f(0)} EN(E)dE = \int_0^{\infty} f(E)ECE^{\frac{1}{2}}dE \quad \text{where C is the constant depending over other}$$

properties.

And at absolute zero f(E) = 1 so

$$U(0) = C\int_0^{E_f(0)} E^{\frac{3}{2}}dE = C\frac{2}{5}E_f^{\frac{5}{2}}(0)$$

$$= \frac{3}{5}NE_f(0)$$

So $\overline{E} = \dfrac{3}{5}E_f(0)$ hence proved

5. Calculate the Fermi energy in copper. Consider density of copper as 8.94 x 10^3 kg/m^3 with atomic mass 63.5 amu.

Solution:

Here electron density = N/V atom/m^3

$$\frac{N}{V} = \frac{N_A\rho}{W} = \frac{6.02\times10^{26}\times8.94\times10^3}{63.5}$$

$$= 8.48 \times 10^{28} \text{ atoms/m}^3$$

Fermi energy $E_f = \dfrac{h^2}{2m}\left(\dfrac{3\pi^2 N}{V}\right)^{\frac{2}{3}}$

$$= \frac{\left(1.05\times10^{-34}\right)^2}{2\times9.1\times10^{-31}}\times\left(3\times9.86\times8.48\times10^{28}\right)^{\frac{2}{3}}$$

$$= 0.06\times10^{-37}\left(2508.38\times10^{27}\right)^{\frac{2}{3}}$$

$= 6.91 \, eV$

6. Table shows the results for Fermi energies of some monovalent elements

Metal	Cu	Li	Rb	Cs	Ag	K
E_f	7.04	4.72	1.82	1.53	5.51	2.12

If the Fermi velocity of electron in one of the metals is 0.73×10^6 m/s. Identify the metal. Also calculate the Fermi temperature.

Solution:

At Fermi energy,

Kinetic Energy $= E_f = \dfrac{1}{2}mv_f^2 = 0.5 \times (9.1 \times 10^{-31})(0.73 \times 10^6)^2$

$= 2.42 \times 10^{-19}$ joules

$= 1.51 \, eV$ hence the metal is Cesium.

And Fermi energy at Fermi temperature, $E_f = kT_f$

Or $T_f = \dfrac{E_f}{k} = \dfrac{2.42 \times 10^{-19}}{1.38 \times 10^{-23}}$

$= 1.75 \times 10^4$ K.

7. If the Fermi energy of any metal is 10 eV. What is the corresponding classical temperature?

Solution:

We have $E = \dfrac{3}{2}kT = \dfrac{3}{5}E_f$

So $T = \dfrac{2E_f}{5k} = \dfrac{2 \times 10 \times 1.6 \times 10^{-19}}{5 \times 1.38 \times 10^{-23}}$

$= 4.64 \times 10^4$ K.

8. There are about 25×10^{28} free electrons/m^3 in sodium. Calculate its Fermi energy, Fermi velocity and Fermi temperature.

Solution:

Here $\dfrac{N}{V}$ = electron density $= 2.5 \times 10^{28}$ /m^3

Now Fermi energy $E_f = \dfrac{h^2}{2m}\left(\dfrac{3N}{8\pi V}\right)^{\frac{2}{3}} = \dfrac{\left(6.62\times10^{-34}\right)^2}{2\times9.1\times10^{-31}}\left(\dfrac{3}{8\pi}\times2.5\times10^{28}\right)^{\frac{2}{3}}$

$= 5 \times 10^{-19}$ Joules

$= 3.1$ eV

This is the maximum kinetic energy of free electron at 0 K.

Then $E_f = \dfrac{1}{2}mv^2$ or $v_f = \left(\dfrac{2E_f}{m}\right)^{\frac{1}{2}} = \left(\dfrac{2\times5\times10^{-19}}{9.1\times10^{-31}}\right)^{\frac{1}{2}}$

$= 1.047$ m/s

And $E_f = kT_f$

So $T_f = \dfrac{5\times10^{-19}}{1.38\times10^{-23}}$ $= 3.623 \times 10^4$ K.

9. Calculate the extent of the energy range between f(E) =0.9 and f(E) = 0.1 at 200 K and express the same as a function of E_f which is 3 eV.

Solution:

We know that

$$f(E) = \dfrac{N_i}{g_i} = \dfrac{1}{1+e^{\left(E_i - E_f\right)/kT}}$$

$$= \dfrac{1}{1+e^{-\left(E_f - E\right)/kT}} = \dfrac{1}{1+e^{-x}}$$

Where $x = \dfrac{E_f - E}{kT}$

(i)   When f(E) = 0.9

so $0.9 = \dfrac{1}{1+e^{-x}} \Rightarrow 0.9e^{-x} = 1-0.9 = 0.1$

$e^{-x} = \dfrac{1}{9}$ or $e^x = 9$

So $x = \log_e 9 = 2.198$

$\boxed{E_f - E = 2.198 \times kT}$

Now $kT = \dfrac{1.38\times10^{-23}\times200}{1.6\times10^{-19}}$ eV

$= 0.017$ eV

(ii)   When f(E) = 0.1

$0.1 = \dfrac{1}{1+e^x}$ on solving $x = 2.3026\log_{10} 9$

$$\frac{E_1 - E_f}{0.017} = 2.3026 \times 0.954$$

$E_1 = E_f + 0.0.37 = 3 + 0.037 = 3.037$ eV

So $\Delta E = E_1 - E = 3.037 - 2.963 = 0.074$ eV

Or $\dfrac{\Delta E}{E_f} = \dfrac{0.074}{3} = 0.025 = 2.5\%$

10. Calculate the temperature at which there is only one percent probability that a state, with energy 0.5 eV above Fermi energy will be occupied by an electron.

Solution:

We know that $f(E) = \dfrac{1}{1 + e^{(E_i - E_f)/kT}}$

Here $E - E_f = 0.5$ eV and $f(E) = 1\% = 1/100$

So $\dfrac{1}{100} = \dfrac{1}{1 + e^x}$

So $e^x = \dfrac{0.99}{0.01} = 99$

Or $x = 2.3026 \log_{10} 99$ or $kT = \dfrac{0.5}{2.3026 \log_{10} 99} = 1.109$ eV

Or $T = \dfrac{1.109 \times 1.6 \times 10^{-19}}{1.38 \times 10^{-23}} = 1264$ K

SUMMARY

- Phase space is required to represent a state of the entire system. Any point represents position and state of motion.
- The volume of a unit cell in phase space is h^3.
- Microstates are the internal arrangements of the system whereas macrostate are macroscopic arrangements of the system.
- Entropy of the system is related to thermodynamic probability as $S = k_b \ln W$
- A group of systems or assembly of systems is called an Ensemble.
- In MB statistics, particles are identical, distinguishable and do not obey Pauli's exclusion principle, without spin and total probability of distribution is given by

$$W = \lfloor N \frac{(g_i)^{N_i}}{\lfloor N_i}$$

- In BE statistics, particles are identical and indistinguishable and do not obey Pauli's exclusion principle, with integral spin and here

$$W = \prod_{i=1}^{n} W_i = \prod_{i=1}^{n} \frac{\lfloor N_i + g_i - 1}{\lfloor N_i \lfloor g_i - 1}$$

And $$f(E) = \frac{N_i}{g_i} = \frac{1}{e^{(E_i - \mu)/kT} - 1}$$

- In FD statistics, particles are identical, indistinguishable, obey pauli's exclusion principle and have half integral spin and here

$$W = \prod_{i} \frac{g_i}{\lfloor N_i \lfloor g_i - N_i}$$

And $$f(E) = \frac{N_i}{g_i} = \frac{1}{1 + e^{(E_i - E_f)/kT}}$$

If T>0 f(E) = ½

If T=0 and $E_f > E_i$ f(E)=1

And if $E_f < E_i$ f(E)=0

- Total number of particles $$N - \frac{8\pi V}{3} \left(\frac{2mE_f}{h^2} \right)^{\frac{3}{2}}$$

   And Fermi Energy $$E_f = \frac{h^2}{2m} \left(\frac{3n}{8\pi} \right)^{\frac{2}{3}}$$

- Average energy $$\overline{E} = \frac{3}{5} E_f$$

- Fermi Velocity $$v_f = \frac{h}{m} \left(\frac{3n}{8\pi} \right)^{\frac{1}{3}}$$

- And average velocity $$\overline{v} = \frac{3}{4} v_f$$

Comparison between MB, BE and FD statistics

S.no	Type	MB	BE	FD
1	Nature of particles	Distinguishable, Boltzons	Indistinguishable, Bosons	Indistinguishable, Fermions
2	Spin	-	0 or integral $(0,1,2...)$	Half integral $(1/2, 3/2, 5/2...)$
3	No. of Particles per energy state	No upper limit	Do not obey Pauli's exclusion principle	Obey Pauli's exclusion principle, only one particle per energy state
4	$f(E) = \dfrac{N_i}{g_i}$	$\dfrac{1}{e^{\alpha + \beta E_i}}$	$\dfrac{1}{e^{\alpha + \beta E_i} - 1}$ where $\alpha = -\dfrac{\mu}{kT}$ and $\beta = \dfrac{1}{kT}$	$\dfrac{1}{e^{\alpha + \beta E_i} + 1}$ where $\alpha = -\dfrac{E_f}{kT}$ and $\beta = \dfrac{1}{kT}$
5	Nature of application	Applicable to ideal gases	To photons, phonons with symmetric wave function	To electron, proton with anti-symmetric wave functions

EXERCISES

1. What do you understand by (i) phase space (ii) Microstate (iii) Macrostate?
2. What are fermions, Bosons and Boltzons? Give their examples.
3. Define thermodynamic probability. Give relation between Entropy and thermodynamic probability.
4. Give the sketch of Fermi distribution for T=0 and T >0 K and explain it.
5. Give characteristics of MB, BE and FD statistics. Give their limitations also.
6. Explain Fermi Energy. Prove that average energy is $\dfrac{3}{5}E_f$ at absolute zero.
7. Calculate the number of energy states in energy range E and E+dE in phase space and also find the number of states in momentum range p and p+dp.
8. Give postulates of MB, BE and FD statistics.
9. Discuss the requirements of quantum statistics.
10. Write down probability distribution functions for MB. BE and FD statistics.
11. Deduce the expression of Fermi energy at absolute zero and prove that it depends only on electron concentration and independent of the size of the conductor.
12. Give expression for BE statistics and derive Planck's blackbody radiation law.
13. Three distinguishable particles each can be in E, 2E, 3E and 4E energy state with total energy 6E. Give estimate of total possible distribution of particles in energy states. Find also the microstates in each case. (WBUT 2007).
14. Consider a two particle system, each of which can exist in a state E1, E2, E3. What are the possible states if the particles are (i) Bosons and (ii) Fermions. (WBUT 2006).

15. Three identical particles can be in any of the four states. What are the number of possible ways of distributing them in various states according to MB, BE and FD statistics.

16. If the Fermi velocity of the electron in a metal is 0.86 x 10^6m/s then find Fermi energy and Fermi temperature.

17. Show that the wavelength associated with an electron having energy equal to the Fermi energy is given by $\lambda_f = \left(\dfrac{8\pi V}{3N}\right)^{\frac{1}{3}}$. if this wavelength is 5.2 nm then determine the Fermi temperature.

18. Determine the number of energy states available for the electrons in a cubical box of 2 cm side below energy of 3 eV.

19. Determine the temperature at which there is one percent probability that a state with an energy 0.45 eV above the Fermi energy, will be occupied by an electron.

20. Find the electron concentration of silver atom with atomic weight 108 and number of free electron per atom as one and Fermi energy 4.5 eV at 0 K.

....

www.ingramcontent.com/pod-product-compliance
Lightning Source LLC
Chambersburg PA
CBHW051211170526
45166CB00005B/1852